PREDICTING INTERCITY FREIGHT FLOWS

TOPICS IN TRANSPORTATION

Series Editor
M. Florian, *Centre de Recherche sur les Transports, Montréal, Canada*
Editorial Board
R. Allsop, *University College London, UK*
M. Ben-Akiva, *MIT, USA*
S. Dafermos, *Brown University, USA*
E. Hauer, *University of Toronto, Canada*
M. Koshi, *University of Tokyo, Japan*
W. Leutzbach, *University of Karlsrühe, FRG*
A. May, *University of Leeds, UK*
E. Morlok, *University of Pennsylvania, USA*
D. Robertson, *TRRL, UK*

Also in this series

Information Technology Applications in Transport
Peter Bonsall and Michael Bell

Books of Related Interest

New Survey Methods in Transport
2nd International Conference,
Hungerford Hill, Australia, 1983

Proceedings of the Ninth International Symposium
on Transportation and Traffic Theory,
Delft, 1984

Behavioural Research for Transport Policy:
The 1985 International Conference on Travel Behaviour,
Noordwijk, 1985

PREDICTING INTERCITY FREIGHT FLOWS

Patrick T. Harker
The Wharton School, University of Pennsylvania, USA

VNU SCIENCE PRESS
Utrecht, The Netherlands

VNU Science Press BV
P.O. Box 2093
3500 GB Utrecht
The Netherlands

First published 1987

CIP-DATA KONINKLIJKE BIBLIOTHEEK, DEN HAAG

Predicting intercity freight flows / Patrick
T. Harker. — Utrecht : VNU Science Press.
Ill. — (Topics in transportation)
With index, ref.
ISBN 90–6764–064–6 bound
SISO 375.3 UDC 656.073
Subject heading: transportation.

Printed in Great Britain by J. W. Arrowsmith Ltd., Bristol

TO EMILY

CONTENTS

PREFACE

The field of transportation science has a long tradition in the analytical modelling of urban transportation systems. In particular, the prediction of equilibrium flows and origin–destination demand patterns is at the heart of much of the recent theoretical literature in transportation. The modelling of the intercity freight transportation system, however, has not been subjected to the same degree of theoretical and mathematical rigor until very recently. Given this system's increasing importance throughout the world, these types of sophisticated models are vital in aiding planners in both the public and private sectors.

This monograph is a revised version of my Ph.D. thesis which was written at the University of Pennsylvania and Argonne National Laboratory in 1982–83. It presents a general model of the intercity freight transportation system which should prove to be a useful tool, as it stands, as well as being a springboard for several extensions. There are several differences between this volume and my earlier dissertation. First, the notation throughout this monograph has been greatly altered. Second, the development of the class of models in which the Generalized Spatial Price Equilibrium Model is the most general follows along the lines of the two recent papers by myself and Terry L. Friesz which appeared in *Transportation Research* **19B** (1985). Third, the chapter on solution algorithms has been expanded to include a discussion on several alternative algorithms to those used in this study; and, finally, Chapter 7 has been added in order to include a discussion of the implications of recent work in imperfect commodity markets on the model developed herein.

Much of the material in Chapters 2, 3 and 6 has appeared in the literature. Specifically, the material in these chapters has been adapted from the following papers:

P.T. Harker (1985). The state of the art in the predictive analysis of freight transportation systems, *Transp. Rev.* **5**(2), 143–164.

P. T. Harker and T. L. Friesz (1985). Prediction of intercity freight flows, I: theory, and II: mathematical formulations, *Trans. Res.* **19B,** forthcoming.

P. T. Harker and T. L. Friesz (1985). The use of equilibrium network models in logistics management: with application to the U.S. coal industry, *Trans. Res.* **19B**(5), 457–470.

In addition, the material in Section 5.4 and Chapter 7 can be found in much greater detail in a series of working papers in the Department of Decision Sciences, The Wharton School, University of Pennsylvania.

The initial suggestion to turn my dissertation into a book was made by the Senior Editor, Professor Michael Florian of the University of Montréal. I am indebted to my colleague and thesis advisor, Professor Terry L. Friesz, for all of his help and encouragement. Other individuals who gave invaluable assistance include: Professor Edward K. Morlok, Professor Tony Smith, Dr. Roger L. Tobin, and Dr. Joel Gottfried. Financial support for the research presented in this book has come from a National Science Foundation Graduate Fellowship, a fellowship from Argonne National Laboratory, a grant from the University of California, Santa Barbara, and National Science Foundation Research Initiation Grant CEE–840392.

Finally, as I stated in my dissertation, I could never have accomplished this work without the love and support of my wife, Emily.

Patrick T. Harker
Philadelphia, Pennsylvania USA
February 1986

Chapter 1
INTRODUCTION

1. INTRODUCTION

The transportation industry is a vital part of a nation's economy. Freight transportation not only affects the availability and prices of goods sold at market, but it also has a major impact on energy usage, national defense matters, and many other national concerns. An inadequate supply of transportation service constitutes a "bottleneck" in the economy, for without an adequate transportation system, the growth of a nation's economy is limited at best. Thus, the study of this industry and its impacts on other sectors of our society is an important component in understanding our complex world.

Due to the complexity of the freight transportation system, there has developed an enormous diversity of research focused on this area. From the detailed scheduling of rail cars to the impacts of freight movements on national energy usage, reseachers have attempted to describe how freight transportation affects and is affected by physical and economic conditions. By understanding and improving this system, we are improving the overall health of the economy.

This monograph focuses on the issue of predicting the usage of freight transportation services and the production of these services over the transportation network. The uses of such predictions are many. Governmental agencies, studying various policy alternatives, can use this type of approach to predict the effect which their policies would have on the economy via the response of the transportation sector. Likewise, private firms may wish to know how their business decisions would affect their competitive stance, and thus would want to predict the responses of the transportation sector to their decisions. Therefore, the goal of this monograph is to develop a model of the freight transportation system which adequately describes this system and its relation to the overall economy so that the model can be used for policy analysis.

Ideally, such a model would incorporate all the complex interactions in the system. However, this is not a task which can be done overnight, but takes years of research and many minds working to create a body of theory about this complex subject. This monograph presents the first simultaneous treatment of the producers of goods, the consumers of goods, the producers of the transportation services (the carriers) and the consumers of these services (the shippers). The rationale behind the building of this model is to create a theoretical foundation on which to expand our knowledge of the freight transportation system. As Saaty and Alexander (1981, p. 11) write:

> Modeling involves a heroic simplication of a problem using the minimum possible number of basic variables in order to come to grips with the essentials. The first attempt usually comprises a stepping stone to more sophisticated elaborations of the model. To build an edifice, one requires a well-planned foundation, without which there could be no sound structure.

Therefore, the purpose of this monograph is to be the stepping stone towards formally treating the generation, distribution, modal split and assignment of freight movements in a consistent, general equilibrium framework. In this model, the decisions of all agents involved in freight movements, both those who wish to have goods moved (the shippers) and the transportation firms (the carriers) are treated simultaneously. Both the theoretical and computational aspects of such a model are developed, and then applied to the United States coal economy.

The remainder of this monograph is divided into seven chapters. In Chapter 2 a typology of the various approaches to the modeling of the freight transportation system is developed, and the major works in each area are discussed. As in any new work, we must build from the past, and this work is no different. As Weintraub (1979, p. 160) writes:

> Economic [and all other] knowledge is constructed piece by piece, using partial explanations which are better fitted into the nooks and crannies of our discipline than other partial explanations.

Chapter 2 critically evaluates the previous work in the freight area and synthesizes the major concepts of this literature.

These concepts are then used as the basis to develop a generalized spatial price equilibrium model of the freight transportation system in Chapter 3. The conceptual framework of this model is developed, two equivalent mathematical statements of the problem are made, and the questions of the existence and uniqueness of a solution of this model are addressed.

In Chapter 4, the theoretical and econometric approaches to the question of what rate carriers will charge for their services are explored. In particular, the cases of purely competitive and contestable markets are treated at length.

Solution algorithms for this model are then developed in Chapter 5. The first is based on the nonlinear complementarity form of the model and the second on the model's variational inequality representation. The special case of rates equal to marginal costs, which arises from purely competitive and perfectly contestable markets, yields an efficient diagonalization/Frank-Wolfe-type algorithm, and is also discussed in this chapter. A series of examples are used to test the solution algorithms presented.

Chapter 6 presents the results of the application of this model to the U.S. coal economy. The model is run and compared against base year (1980) data, and then run to show its usage as a policy analysis tool in a scenario where certain U.S. ports are closed and exports double in volume. Recent advances in spatial price equilibrium theory are discussed in Chapter 7, and it is shown how these new extensions of the commodity market model can be integrated into the freight system model which was developed in Chapter 3.

Finally, Chapter 8 discusses the contributions of this work and sketches out the future research needs in the area of predictive modeling of the freight transportation system.

Chapter 2
PREDICTING INTERCITY FREIGHT FLOWS: THE STATE OF THE ART

The complexity inherent in modeling the freight transportation system has led researchers down many varying paths in the pursuit of insights into and understanding of this system. As Chiang, et al. (1980, p. 13) write:

> Models have been developed by researchers from many disciplines using many different approaches in an attempt to solve many different problems. This is just one indication that freight transportation involves a complicated decision-making process.

The purpose of this chapter is to explore the basic approaches which researchers have taken, highlighting the major works in each category, pointing out their achievements and shortcomings. This review will allow us to critically evaluate the model developed in Chapter 3, pointing out what features of the previous methodological approaches this model contains and which features it ignores. Let us begin with a definition of the problem at hand, namely, predictive analysis of freight transportation systems.

2.1 Conceptual Model of an Intercity Freight Transportation System

The concept of predictive analysis of freight transportation systems is difficult to define due to the enormous diversity of the types of predictions made about this system. Models have been developed which range from predicting detailed routing of freight to the prediction of entry and exit of firms from the transportation industry. However, all of these models have at their core the concept of forecasting the behavior of the agents involved in the movement of freight. To begin our discussion, let us define the set of agents which are commonly considered in modeling the freight transportation system.

Figure 2.1 depicts the set of agents which we will consider as comprising the freight transportation system and the interrelations between them. Let us explore this figure in greater detail.

The producers are those economic agents whose role in this system is the production of goods. The consumers are those agents who consume these goods. Typically, a subset of producers and a subset of consumers are defined to act in a certain subregion of the total region under study. The economic force by which these two groups of agents 'communicate' is the set of market prices of goods which they are selling and buying.

Since we have defined the producers and consumers to reside and act in various subregions, there must be some economic agent whose role is to coordinate movements between the various regions. The shippers are that set of economic agents who make the decisions on the generation of trips from an origin, the distribution of these trips to the set of destinations, and the set of transportation firms who will move the freight from the origin to the destinations. The shippers are actually a conglomeration of various decision-making entities, such as shipping departments of manufacturing firms, distribution departments, freight-forwarders, receiving departments of firms, etc. Depending upon the purpose of the model, shippers may actually be defined as one or more of these entities. In general we only wish to consider the shippers as the set of agents who make the decisions on the movement of freight in the region under study. These decisions are based upon the market conditions, which we will say are summarized via the market prices. As Roberts (1977, p. 3) writes, "... the only motivation for moving freight is an economic one." Therefore, the shippers' choice to move freight depends upon the supply and demand behavior of the producers and consumers respectively, and the market prices associated with this behavior.

As stated above, one of the shippers' roles is to decide by what means the freight will be moved, the means being the set of

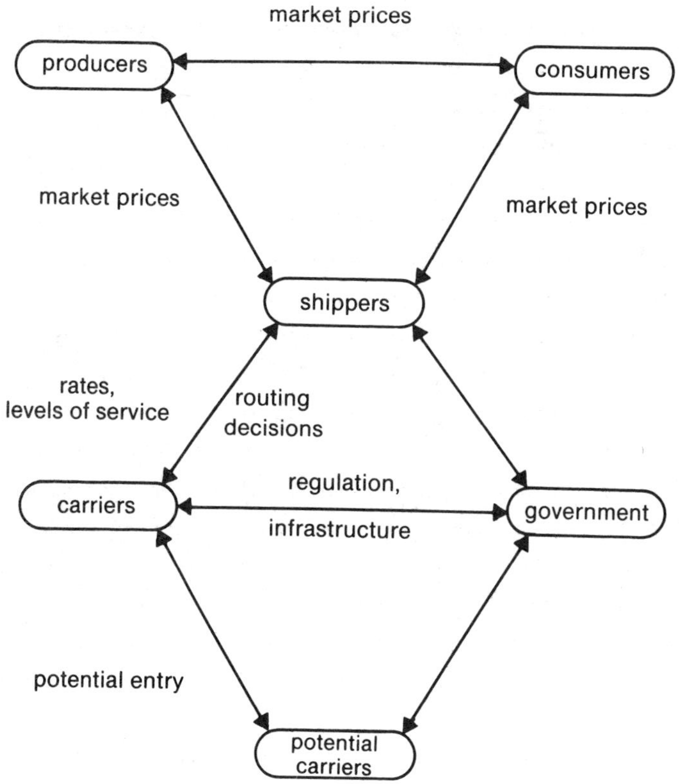

Figure 2.1: Relationship Among Agents

transportation firms which we shall call the carriers. In general, the set of carriers includes all the various modes of freight movement. Also, a common assumption is that the carriers behave as profit maximizing firms. Therefore, the carriers are defined as profit maximizing firms who produce movements of freight as their outputs.

The relationship between the shippers and carriers is one of consumers and producers of the transportation service. By their choice of the carrier or set of carriers to which they wish to give the movement of a load of freight, which we will call the shipper's routing decision, the shippers create a demand for the carriers' outputs. The carriers, on the other hand, will charge a rate for this movement and in producing it, will create a certain level of service associated with this movement. This relationship between shippers and carriers is shown in Figure 2.1 by a two-way arrow.

There are two other agents which we shall consider as part of the freight transportation system. Potential carriers are economic agents who do not currently offer any transportation services in the market, but have the potential to do so. They are important in discussing the issue of freight rates due to the pressure their potential entry places on the currently operating carriers.

The government is defined as the set of federal, state and local agencies involved in any way with freight transportation. The two major ways in which the government enters this system are regulation and the provision of transportation infrastructure. Regulatory policies affect the decisions of both the shippers and carriers, and the entry into the market of the potential carriers. By providing a large portion of the freight system's infrastructure, i.e., the highways and waterways, the government also becomes involved witha the operations of the carriers.

Given this conceptual framework in which we shall discuss the problem of predictive analysis of the freight transportation system, we shall now turn to a review of the three general approaches used for this problem: the econometric model, the spatial price equilibrium model, and the freight network equilibrium model. Let us begin with a review of the econometric modeling approach.

2.2 Econometric Models

The econometric modeling approach to freight transportation systems analysis involves the use of time series and/or cross-sectional data to estimate structural relationships which describe the behavior of a part of or all of this system. Data on the 'materials' which are used to produce transportation services (labor, capital, energy, etc.) are typically utilized to estimate industry-wide or firm-specific cost or production functions. On the demand side, data on rates, level of service attributes and demands for transportation services are used to estimate demand functions for this service. Typically, models which we shall call econometric do not consider a detailed description of the transportation network. That is, econometric models rely on very simple descriptions of the network, ignoring for the most part the complexities of an actual transportation system. One reason for this type of treatment of the transportation network is that there is little data available to estimate economic relationships on the actual network. The explicit treatment of the complexities of the actual transportation network falls into the realm of what we shall call network models.

The econometric modeling approach typically focuses only on the shipper-carrier-government relationship, ignoring the other agents and interrelationships shown in Figure 2.1. Also, this type of modeling approach has not been used for and does not lend itself to use in asking questions about the detailed routing of freight since it is difficult to incorporate a detailed network representation, as discussed above. This routing question falls into the realm of the network models which we will discuss in the next two sections.

The econometric model does have some advantages over the network modeling approach. First, by often working from the basic 'materials' of production, the impacts of various policies on labor, capital, etc., are easily studied. Second, econometric models allow for variation in these basic production components, and thus are easy to use in the study of the entry and exist of firms in the transportation industry and other studies which involve dynamic adjustments.

The work done in the areas of econometric modeling of freight transportation can be classified into one of three categories: the supply-side models, the demand-side models, and the integrated models, encompassing both supply and demand behavior. In what follows, we will define and briefly review the major works and conclusions of each category. Let us begin with the supply-side models.

The supply-side models focus on the issue of describing the production of freight transportation services. The major impetus for the development of these models was not to make predictions about the freight transportation system, but rather to understand the production/cost characteristics of the industry. The results of these models were used in the analysis of the regulatory reform of this industry. These models, though not meant for direct use as predictive tools, are useful in developing such tools in that they shed light upon the definition of potential equilibrium industry structures and output vectors. Therefore, it is instructive to briefly review the findings of the supply-side models.

The study of the cost characteristics of the railroad industry has a long history. The questions of whether there are economies of scale (costs decreasing as the output of a firm increases) and/or economies of route density (costs on a particular route decreasing as traffic on that route increases) have played an important role in regulatory reform debates. The works by Klein (1947), Meyer, et al. (1959), Borts (1952, 1960), and Healy (1961, 1962) are inconclusive in their answers to the above questions, as Keeler (1983, p. 51) points out. These studies found few economies of scale and density.

There were, however, methodological problems associated with the above studies. The problems mainly centered on the distinction between economies of scale and density, the allocation of costs between freight and passenger movements, and the use of ton-miles (or passenger-miles) as the sole measure of output. The studies by Keeler (1974), Griliches (1972), Hasenkamp (1976), Harris (1977), Harmatuck (1979), Brown, et al. (1975), Caves, et al. (1981), McFarland (1978) and Friedlaender and Spady (1981) all attempted to overcome these problems by using alternative functional forms and/or alternative definitions of output. These studies all find strong evidence for economies of density, but are inconclusive in their findings on the economies of scale (see Keeler, 1983, p. 51).

Jara-Diaz (1981) has taken a different approach from the above studies by looking at the problem of transportation cost estimation from a multi-product viewpoint. He treats each origin-destination (O-D) move as a separate output and then applies the multi-product cost function concepts explained in Baumol, Panzar and Willig (1982). Both economies of scale and economies of scope (costs decrease as more types of outputs are produced--more O-D pairs served) are observed in his sample of a Class III railroad.

Another innovation in modeling railroad costs is the 'hybrid' cost function approach by Braeutigam, Daughety and Turnquist (1982), Daughety and Turnquist (1980) and Daughety, et al. (1983). In this approach, engineering cost estimates derived from a network representation of the 'production' function are merged with other firm-wide data to form what they call a 'hybrid' cost function.

In summary, there appears to be strong economies of density in railroad operations. Economies of scale are also present, but are weaker than the economies of density. Future research along the lines of that by Jara-Diaz and Daughety, et al. may yield further insights into the issue of railroad costs (see Keeler, 1983, Chapter 3 and Appendix B for a further discussion of this topic).

There has been relatively less work done in the area of motor carrier, barge and pipeline costs. In the trucking industry, the early studies by Roberts (1956) and Nelson (1956) find limited or no economies of scale in this industry. Later studies by Warner (1965), Ladenson and Stoga (1974), Dailey (1973) and Lawrence (1976) are inconclusive in their assessment of the economies of scale issue. In more recent work, both Friedlaender (1978) and Chow (1978) find diseconomies in the long-haul less-than-truckload (LTL) and truckload (TL) industries, while Chow finds some economies of scale in the short-haul LTL industry. In summary, it appears that there is not strong evidence to support the claim that economies of scale exist in the motor carrier industry.

In the inland waterway industry, the studies by Case and Lave (1970) and Polak and Koshal (1976) both find increasing returns to scale. In the pipeline industry, Cookenboo (1955) and Hazard (1977) also show economies of scale. However, these studies are somewhat dated in their methodologies, and thus it is difficult to make definitive statements on the economies of scale issue in these industries.

The demand-side models attempt to explain the demand for transportation service as a function of the rate charged for this service plus the level of service (LOS) offered. These models are useful to review in that they elucidate which characteristics of freight movements should be incorporated into predictive models of the freight system. Therefore, let us briefly review the major developments in understanding the demand for freight transportation services.

The groundwork for the models of freight transportation demand is the Lancaster (1966) activity approach to consumer behavior in which it is claimed that consumers derived utility from the attributes of a good, not the good itself. Quandt and Baumol (1966) use this idea in the development of the abstract-mode concept. In this concept, modes are defined by a vector of attributes, such as reliability, price, etc. Therefore, the modes are abstract in the sense that it is the attributes, not the technology used, which defines a mode.

Following Winston (1983), we may classify the demand models as aggregate or disaggregate. In the disaggregate category, there are the behavioral models and the inventory-theoretic models. Let us begin with the aggregate models.

The two major aggregate models are those by Oum (1979) and the demand study in Friedlaender and Spady (1981). These models assume that the producing firms are profit maximizers and that transportation is a factor in their production processes. Assuming perfect competition in the factor (transportation) market, the demand for transportation can be found via Shepard's Lemma (see Varian, 1978, p. 32). The demand function for transportation thus relies on the assumed cost function of the producing firm.

In order to better understand the derived nature of transportation demand, researchers began to look at the decisions made by a single firm, thus creating models in the disaggregate category. The behavioral models look at the decision processes of the agents involved in making the shipping decisions. Allen (1977) takes this approach, developing a model of freight transportation demand consistent with the profit maximizing behavior of the firm.

Daughety and Inaba (1978) then extended Allen's ideas by adding uncertainty to the profit maximization problem for the firm. This uncertainty arises out of the stochastic nature of service attributes and from unknown decision attributes. This approach leads to a logit demand formulation which is used by Daughety (1979), Daughety and Inaba (1981) and Levin (1981).

Winston (1981) points out that due to risk aversion not being equal over all modes, the assumption of independence of the stochastic error terms necessary to derive the logit formulation may not be realistic. Thus, he includes dependent error distributions and formulates a probit function for freight transportation demand.

The inventory-theoretic approach to freight transportation demand is based upon the perspective that the transportation decisions are being made by the inventory manager of a firm. That is, shipping decisions are modeled as resulting from inventory considerations. Baumol and Vinod (1970) began this approach in earnest, and were followed by the work of Das (1974) and Constable and Whybark (1978).

The work done at MIT on freight demand also falls under this category. Some of the reports detailing this work are Chiang, Roberts and Ben-Akiva (1980), Roberts (1976), and Terzeiv (1976). Finally, shipment size, shipping frequency and other decision attributes have been shown by McFadden and Winston (1981) to have a significant effect when added to a demand model.

More detailed reviews of freight transportation demand models can be found in Smith (1975) and Winston (1983).

Finally, the integrated econometric models which take both a supply and demand model and calculate an equilibrium prediction fall into the category of econometric models. The major works in this area are those of Friedlaender (1969) and Friedlaender and Spady (1981). Building upon the cost and demand models cited previously, they calculated the competitive equilibrium (marginal cost pricing) for the purpose of evaluating regulatory reform.

However, there is an internal inconsistency with this approach. There are scale economies present in the cost functions, and thus there is no reason to believe that a competitive equilibrium will occur in reality. Recognizing this, Friedlaender and Spady (1981) justify this approach by saying that a competitive equilibrium is a useful benchmark to use in policy analysis (pp. 74-75).

The type of equilibrium analysis done with econometric models is intended to answer policy issues with broad impacts, e.g., deregulation, and not for detailed routing questions. The latter question is best answered via a network model. Let us now turn to a review of the network models used in analyzing the freight system, starting with the spatial price equilibrium models.

2.3 Spatial Price Equilibrium Models

In both the spatial price equilibrium models discussed in this section and in the freight network equilibrium models presented in the next section, the transportation system is explicitly represented by a network. Therefore, let us briefly discuss some general characteristics of modeling the freight transportation system via the network approach.

A network model describes the transportation system as a set of nodes and arcs representing the system infrastructure. Nodes represent facilities such as terminals, ports, rail yards, etc. Arcs represent highways, rail lines, waterways, etc. In general, some measure of costs and levels of service attributes are associated with each element (node or arc) of the network. This measure may either be a constant or a function of the flows on the network.

A predictive network model implicitly assumes that a short-run analysis will be performed due to the fact that the capital which comprises the network is not typically altered in the analysis done with this type of model. It is true that the network structure is altered in normative network design models, but this approach has never been applied to a predictive analysis due to its computational complexity. Therefore, it is very difficult to use a network-type model to address issues such as entry and exit from the transportation industry, dynamic adjustments, etc. Also, as discussed in the introduction to Section 2.2, network models cannot easily address the issue of the substitutability of capital, labor and energy; this issue lies in the realm of the econometric model.

The spatial price equilibrium model takes this network approach in looking at the interactions of the producers, consumers, and shippers, as shown by the top triangle in Figure 2.1. That is, the interactions of the carriers, the producers of the transportation service, are left out of the analysis in this type of approach. Instead of the carriers, cost functions are defined on the elements of the network to represent the transportation firms. Therefore, the spatial price equilibrium approach does not explicitly deal with the decision processes of the carriers.

The spatial price equilibrium model consists of a subset of the nodes in the network being designated as producing or consuming regions for various commodities, or as both. Links connect these regions directly or through a series of transshipment nodes, where commodities are neither produced nor consumed. Demand functions are associated with each consuming region, and supply functions with each producing region. The shippers 'barter' between these regions until an equilibrium is reached. This equilibrium is characterized by the following two conditions:

(a) if there is a flow of commodity i from region A to region B, then the price in A for commodity i plus the transportation cost from A to B will equal the price of commodity i in B;

(b) if the price of commodity i in A plus the transportation costs from A to B is greater than the price of commodity i in B, then there will be no flow from A to B.

Therefore, the demands for transportation are <u>derived</u> from the market forces across regions, the demands being a result of the equilibration process.

The notion of an equilibrium between spatially separated markets and the resulting transportation demands is an old idea. Cournot (1838) describes the equilibrium of prices as goods move between New York and London. Enke (1951) notices that what Cournot is describing is a network flow problem, similar to the current flows in electrical networks to which he was accustomed. The network structure he considers is a bipartite graph with no transshipment nodes. That is, each node is either an origin or destination, but not both. Figure 2.2 depicts this type of network, which the reader may recognize from a simple Koopmans-Hitchcock transportation problem.

Samuelson (1952), in his much referred to paper, noticed that Enke's formulation of the problem can be cast as an extremal problem, analogous to the minimzation of the total power loss in an electrical network. He refers to his objective function as the 'net social payoff', the Kuhn-Tucker analysis of the resulting extremal problem yielding the desired spatial price equilibrium conditions.

In their chapter on 'Some Unsolved Problems,' Beckmann, et al. (1956) state how the spatial price equilibrium model would be very useful in commodity, or freight transportation analysis. Takayama and Judge (1964) (see also Takayama and Judge, 1971) solve this problem by assuming linear supply and demand functions and constant transportation costs to form a quadratic program. Multiple commodities are included in this model and an example is solved using a quadratic programming technique.

The spatial price equilibrium conditions can also be cast into a mathematical form known as a complementarity problem. A complementarity problem (CP) is to find a vector of variables $x = (x_1, x_2, \ldots, x_n)^t$ such that for the vector function $F(x) = (F_1(x), F_2(x), \ldots, F_n(x))^t$ the following holds, where t denotes the transpose operation:

$$\begin{aligned} & F(x)^t \cdot x = 0 \\ & F(x) \geq 0 \\ & x \geq 0. \end{aligned} \qquad (2.1)$$

If the functions $F_i(x)$ are linear, then (2.1) is a linear complementarity problem (LCP); if nonlinear, then (2.1) is a nonlinear complementarity problem (NCP). Takayama and Judge (1970) and Stoecker (1974) noticed that with linear supply and demand functions and linear transportation cost functions, the spatial price equilibrium model becomes an LCP. MacKinnon (1975)

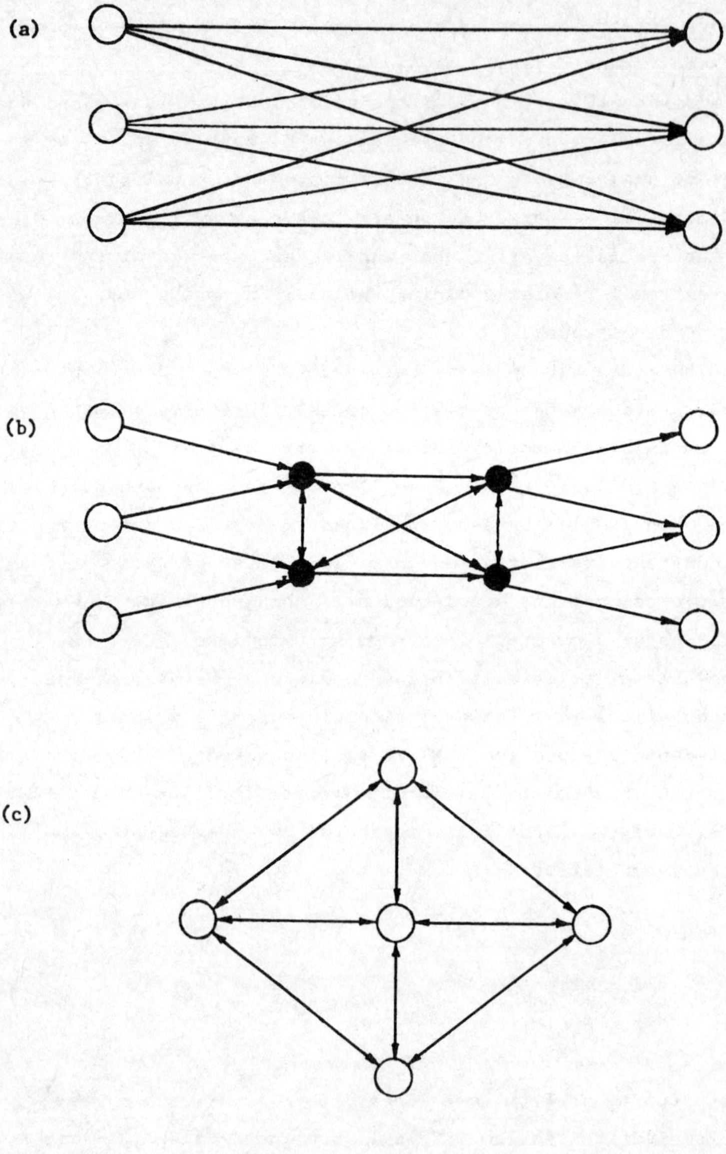

Figure 2.2: Network Topologies Considered in Spatial Price Equilibrium Models

considered nonlinear functions, and thus posed the equilibrium model as an NCP.

Florian and Los (1982) consider the spatial price equilibrium problem on a network with transshipment nodes, as shown in Figure 2.2. They return to the extremal formulation of the problem, utilizing nonlinear functional forms. Their model is stated in terms of path flows between the origins and destinations, thus causing complications in the solution procedures.

Tobin and Friesz (1983) consider a general network, as shown in Figure 2.2, where a node can, in general, be an origin, destination and transshipment point. Also, they show that paths need not be explicitly considered, as in Florian and Los, thus greatly reducing the computational complexity of the model. However, this formulation, as opposed to a path formulation, implicitly assumes there is substitutability of goods at each node in the network. That is, if a node is a pure transshipment point, goods flowing in and out of this node cannot be 'tagged' by their final destinations. When specific origin-destination movements are required, such as in railcar movements, this assumption would not be adequate (see Harker, 1985, for a further discussion of this issue). If this assumption can be used, substantial computational benefits can be obtained. Friesz, et al. (1983) place the above model in an NCP form and examine the questions of the existence and uniqueness of a spatial price equilibrium in this framework. Also, Friesz, et al. discuss the convergence of a complementarity-based solution algorithm for this model.

Many researchers have looked at how to effectively calculate the spatial price equilibrium for the various models mentioned above. MacKinnon (1975, 1976) discusses how to solve his formulation of the problem, and Ahn (1978) studies the computation of spatial price equilibrium in the framework of the PIES (Project Independence Evaluation System) model. Bender's decomposition was applied to the case shown in Figure 2.2 by Polito, McCarl and Morin (1980), while Asmuth, Eaves and Peterson (1979) and Irwin and Yang (1982) discuss the solution of the LCP formulation. Rowse (1981) solves a formulation of the problem similar to Florian and Los (1982) via a nonlinear optimization code named MINOS. Friesz, Tobin and Harker (1981) also use MINOS iteratively in a diagonalization algorithm to solve their arc-formulation of the spatial price equilibrium model. Other work in this area includes that of Fang and Peterson (1980), Peterson (1981), Pang (1981a,b), Pang and Lee (1981), and Friesz, et al. (1984).

The applications of this concept to various commodity studies are numerous. Takayama and Judge (1964b), Plessner (1965), Stoecker (1974), and Nagy, Furtar and Kulshreshta (1980) apply the concept to various agricultural products. The book by Takayama and Judge (1973) contains many other applications in this area. West and Brandow (1964) apply the concept to the dairy industry, Bawden (1966) to international trade, Kennedy (1974) to the world oil market, Waverman (1973) and Brooks (1975) to natural gas, and Uri (1975) to electricity markets.

There has been some work in extending the basic model framework. Stevens (1968) shows how the spatial equilibrium concept can be used in von Thünen location models. Regional income levels are implicitly derived in the model by Thore (1982), and Ponsard (1982) applies some concepts of fuzzy sets to consider partial spatial price equilibria.

2.4 Freight Network Equilibrium Models

The freight network equilibrium approach to the prediction of freight movements is one which utilizes a network structure to represent these movements, the characteristics of this network approach being the same as those discussed in the beginning of Section 2.3. The freight network equilibrium approach focuses mainly on the actions of the shippers, carriers and potential carriers. Gottfried (1983) also discusses the possibility of incorporating the producers and consumers, but to date this incorporation has not been implemented on a large-scale application. Thus, the work which has been done with the freight network equilibrium approach focuses mainly on the interactions depicted in the bottom portion of Figure 2.1.

The first significant multimodal predictive freight network model was by Roberts (1966) and later extended by Kresge and Roberts (1971), this model becoming known as the Harvard-Brookings model. Only the behavior of the shippers was taken into account. Using constant unit costs, each shipper chooses the shortest path for movements from an origin to a destination, the amount moving between an O-D pair being determined by a Koopmans-Hitchcock submodel. The model was applied to the transportation network of Columbia.

Peterson and Fullerton (1975) constructed a predictive network model which incorporates nonlinear cost and delay functions for the elements of the network. Only one commodity is considered, and demands for O-D movements are treated as given. Also, only one decision-making agent is included in the model. They claim that Wardrop's Second Principle of systems optimization, where the total costs on the network are minimized, is preferable to Wardrop's First Principle of user optimization, where agents involved in each O-D move compete noncooperatively for the transportation resources such that they minimize their own costs. If there is one controlling carrier, then systems optimization seems to be a reasonable assumption. However, if the users of the system (the shippers) are to be modeled, then user optimizations seem to be a better assumption. Devarajan (1981) discusses how user equilibrium is a form of the Nash equilibrium to a noncooperative game. Therefore, Peterson and Fullerton seem to be implicitly assuming that they are modeling carrier behavior. The nonlinearities in the model are handled by posing the model as an optimization problem and then applying the network assignment technique of LeBlanc, Morlok and Pierskalla (1975).

Kornhauser (1979, 1982) has developed an interactive model of the freight system in which network cost parameters can be altered in such a way that predicted flows are close to replicating historical flow levels on the network. Constant unit costs are utilized. Although this model is not predictive in the sense that it does not attempt to predict how basic assumptions of behavior are reflected in flow levels, it has proven to be a useful tool for decision-makers. The unique feature of the model is that it contains two submodels, one to address the question of intracarrier routing, and the second for the intercarrier movements. Thus, an explicit treatment of the difference between movements within a carrier's own network and between carriers' networks is made. No model of shippers' behavior is included.

As part of the National Energy Transportation Study (NETS), CACI, Inc. developed a multicommodity freight network model referred to as the Transportation Network Model (TNM) (see CACI, 1980, and Bronzini, 1980, 1982). Shippers' behavior is explicitly modeled, carriers' behavior is not included in the model. The initial purpose of this model was to study the energy efficiency of the various modes of intercity freight movements, and thus some elements of costs which a shipper is not thought of as being concerned with, such as energy costs, are included. Therefore, the decision-maker represented in this model is a type of aggregate between a shipper and a carrier.

Other models which have been developed along the same lines of thought as those above include the model by the U.S. Department of Transportation (see Swerdloff, 1971), the model of coal movement by Chang, et al. (1981), and the work done on western coal movement by Green (1980), Ebeling and Chang (1979) and Ebeling (1981).

Lansdowne (1981) looked at the problem of modeling the shipper-carrier and carrier-carrier interfaces. By studying current industry practices, he defines four principles which can be used to model the shipper-carrier, carrier-carrier, and intracarrier interactions:

1. the only routes which a shipper will choose are those with the minimum number of interlining points,

2. each carrier will use the shortest distance path within his sub-network,
3. of all eligible routes, the one that maximizes the originating carrier's share of the revenue is the one selected, and
4. if there is more than one originating carrier which could serve a particular origin, then the shipment is divided among all these carriers.

These concepts, although not free of problems, as Gottfried (1983) points out, are the first attempt at addressing the question of what the shipper-carrier and carrier-carrier interfaces look like.

In work sponsored by the U.S. Department of Energy involving the staff of Argonne National Laboratory (ANL), Friesz, et al. (1981) developed a predictive freight model called the Freight Network Equilibrium Model (FNEM). The paper by Friesz, Gottfried and Morlok (1981) describes the conceptual framework of FNEM. This model is the first to recognize two distinct groups of agents, shippers and carriers, in a general predictive model of intercity freight movement. The model assumes that shippers act on a perceived network, which is an aggregate of the physical network on which the carriers act. The shippers' behavior on this perceived network is modeled by Wardrop's user optimization principle. Demand behavior was initially treated as a set of fixed O-D demands, and then later extended to elastic (exponential) demand functions. Once the equilibrium on the perceived network is found, the flows on this network are disaggregated to form carrier-specific O-D demands for service. These demands are then routed on each carrier's sub-network by assuming systems optimization on each sub-network. Thus, FNEM is a sequential model, first solving the shipper sub-model, and then passing these results to the carrier sub-models. Gottfried (1983) has applied this model to a national-level mutlimodal, multicommodity example with better predictive results, when compared with historical data, than any other published results.

Charles River Associates (see Egan, 1982) noted that a sequential approach such as FNEM does not guarantee consistency between two sub-models. Therefore, they attempted to iterate between an assignment model and the macro-economic driver which produces an estimate of the generation of trips in a region. Convergence of this procedure was not found, and there is no mathematical proof that there should be convergence in their model. Furthermore, even if convergence is reached, there is no proof that this convergent solution is necessarily the equilibrium point. Thus, this work points out the need to consider the behavior of the various agents in a consistent, simultaneous model.

Safwat (1982) describes in his dissertation an intercity transportation model for Egypt which includes both passenger and freight movements. The generation of trips in a region is incorporated via a linear function relating trips and transportation costs (see also Safwat and Magnanti, 1982). Thus, Safwat attempts to add producers' and consumers' behavior by this linear function, collapsing their decision-process into one known functional relationship.

Harker (1981) (see also Harker and Friesz, 1982) recognizes the inconsistency of the sequential approach used in FNEM, and develops a simultaneous version of this model. Although unable to solve large problems, it is useful as a check on how inconsistent the sequential approach may be.

Friesz, et al. (1985), by assuming the transportation rate equals the marginal cost between an O-D pair, develop a simultaneous model along the lines of FNEM. Variational inequality theory is used to address the questions of the existence and uniqueness of an equilibrium point of such a simultaneous model.

For a more thorough review of this class of predictive models of the freight transportation system, the reader is referred to the recent paper by Friesz, et al. (1983).

2.5 Comparisons and Synthesis of the Three General Approaches

The three general approaches to looking at the problem of prediction in freight transportation systems, although distinct, have many common traits. Figure 2.3 depicts the linkages (solid lines) and potential linkages (dashed line) among the three approaches. Let us explore these relationships.

The spatial price equilibrium (SPE) and freight network equilibrium (FNE) models have a strong tie in that both utilize a detailed network representation of the transportation system. The SPE approach includes the agents in the top triangle of Figure 2.1, while the FNE approach treats the lower half of this diagram, including the shippers. Neither approach is easily applied to studies which concern dynamic market adjustments due to the fixed network infrastructure. That is, the network modeling approach is not amenable to studies in which there is entry and exit of firms, additions/deletions of portions of a firm's network, etc. These types of studies are most easily handled by an econometric model.

The econometric models and the FNE models, with the exception of Gottfried's (1983) work, both do not include an explicit representation of consumer and producer behavior. In both cases, the demand for transportation is given by some type of functional relationship. Thus, the concepts used in each of these approaches are equivalent in the sense that they treat the same set of agents.

There does not currently exist any strong relationship between the SPE approach and the concept of cost/production functions used in the econometric models. However, as will be discussed below, there is a great potential in making this link.

The works of Jara-Diaz (1981, 1982) and that of Braeutigam, Daughety and Turnquist (1982), Daughety and Turnquist (1980), and Daughety, Turnquist and Griesbach (1983) are important contributions in the attempt to integrate the econometric cost/production function with a network representation of the transportation system. Making this link with econometric models strengthens the

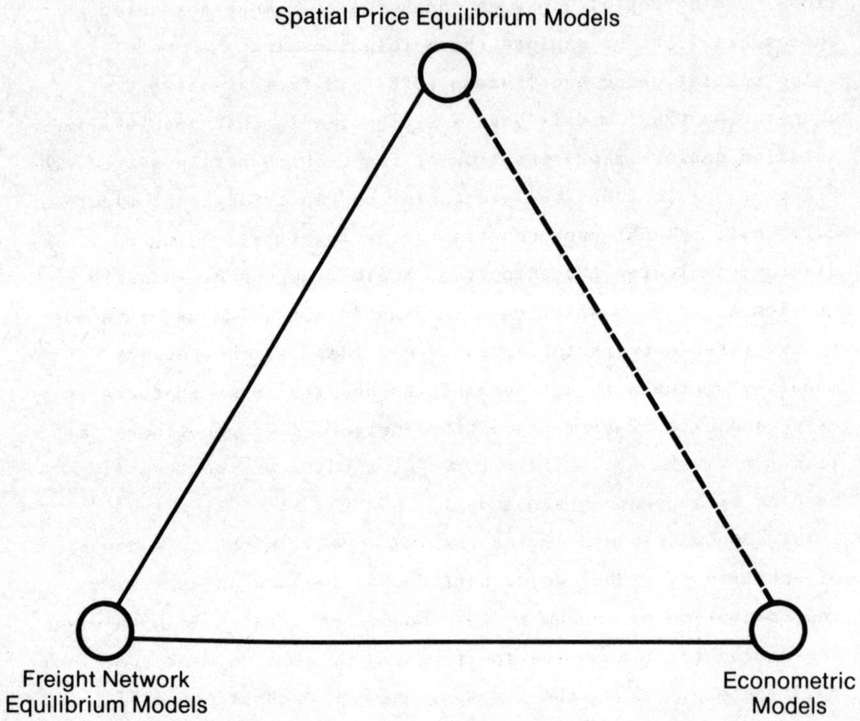

Figure 2.3: Relationships Among the Three General Approaches to Freight System Modeling

overall realism of the predictive analysis. Gottfried's (1983) conceptual extension of his sequential shipper-carrier model to include spatial equilibrium behavior strengthens the tie between the FNE and SPE approaches.

Therefore, the future of the predictive analysis of freight transportation systems seems to be tending toward a strengthening of each node and each arc of the triangle in Figure 2.3. The awareness of the merits of each of these three techniques will hopefully yield stronger, more precise models of this system. The model which is developed in this monograph is a synthesis of the freight network and spatial price equilibrium approaches. This model treats the behavior of the consumers, producers, shippers and carriers in a consistent methodological framework. However, it is recognized that the econometric approaches discussed above must eventually be integrated into this model if an accurate representation of the freight transportation system is to be achieved.

Chapter 3

GENERALIZED SPATIAL PRICE EQUILIBRIUM MODEL

As we have seen in the previous chapter, the current models of the freight transportation system either (a) ignore the details of the network technology on which this system is based -- the econometric models; (b) ignore the role of the transportation firms -- the spatial price equilibrium models; or (c) ignore the importance of the commodity markets in deriving the demands for freight transportation services -- the freight network equilibrium models. The purpose of this chapter is to develop a model -- the Generalized Spatial Price Equilibrium Model or GSPEM -- which attempts to overcome these difficulties by incorporating behavioral models of the producers, consumers, shippers and carriers into a single mathematical statement. Let us begin with a description of the submodels for each agent which will comprise GSPEM.

3.1 Behavioral Models of Agents

Supply-Side. To begin our discussion, let us focus on the production of transport services. This supply side of the transportation market will be assumed to consist of a set K of transportation firms ($k \epsilon K$) or carriers. The outputs which the carriers produce are the set of moves between various origin-destination (O-D) pairs. The level of service offered on an O-D pair is of major importance in transportation decisions, and hence we should associate with each O-D pair a vector of service attributes which the carriers provide at various levels according to their assumed profit-maximizing behavior. However, the inclusion of such a vector will quickly make this model intractable for large problems. Thus, we will approximate the continuum of service levels which a carrier can offer on any O-D pair by a discrete set of service classes; Figure 3.1 illustrates this

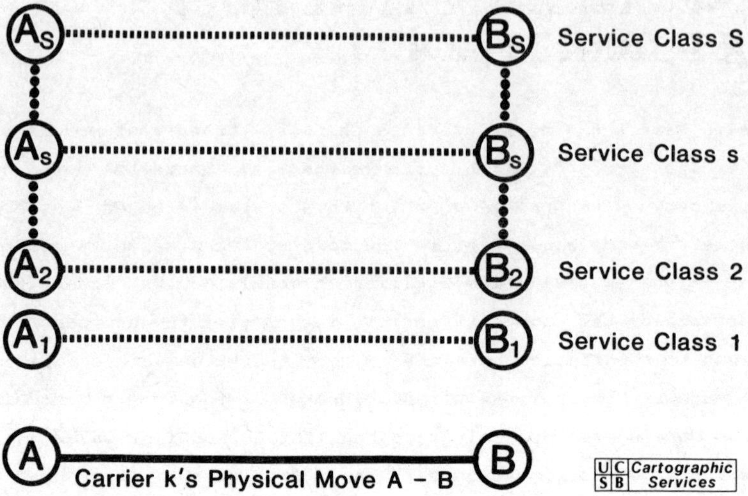

Figure 3.1. Multi-Output Concept

concept. That is, each physical O-D move is represented as a set of O-D moves on an extended network. For example, if carrier k chooses to move a certain quantity of goods between A and B and does so with a level of service s, then this choice is represented in our model by a positive flow on the O-D pair A-B associated with service class s in the extended network. Thus, when we refer to an O-D pair in what follows, we are implicitly referring to both the physical movement and the service level which is being offered; the next section will discuss the representation of service levels in greater detail.

The set K implicitly includes multiple modes of transportation through the definition of each carrier $k \in K$. That is, each carrier is defined to operate a certain mode (railroad, motor carrier, barge, etc.). Competition within and between modes is thus implicitly considered in our discussion of competition between carriers.

The carriers will be assumed to produce their supplies of O-D moves on a fixed network. As Chapter 2 pointed out, the use of such a network implies that a short-run static analysis is to be done since the capital comprising the network (railways, highways, canals) is not altered in this analysis. Therefore, our first assumption is that

(A-1) the economic time frame which is considered in this analysis is the short-run.

To formalize our supply-side description, let N^k and B^k be, respectively, the set of nodes and arcs which carrier k has under his control, and let $H^k(N^k,B^k)$ denote the carrier k's subnetwork. The union of all the carriers' subnetworks consists of the set of $N = \bigcup_{k\epsilon K} N^k$ nodes and the set $B = \bigcup_{k\epsilon K} B^k$ arcs, and is denoted by $H(N,B)$. Let us call this union of all subnetworks the <u>carriers' network</u>. Let us also define

V_k = the set of O-D pairs on carrier k's subnetwork, which are differentiated by both their physical direction and level of service class,

$V = \bigcup_{k\epsilon K} V_k$,

v = an index, $v\epsilon V$

τ_v = the output produced (supplied) on O-D pair $v\epsilon V$

$\tau^k = (\tau_v | v\epsilon V^k)$, and

$\tau = (\tau_v | v\epsilon V)$.

Finally, let us define on each carrier O-D pair v an inverse demand function $R_v(\tau)$ and let us denote the total cost incurred by carrier k in producing τ^k by the function $\underline{C}^k(\tau^k)$.

Two assumptions are made concerning the behavior of the carriers:

(A-2) each carrier is a profit maximizing firm, and

(A-3) the carriers do not collude when setting supply levels.

Therefore, we are assuming that the market consisting of the set K of carriers can be represented by a Cournot or Cournot-Nash quantity model (Chapter 2 of Friedman, 1979). This model assumes that each carrier takes the other carriers' outputs (or strategies) 'as given' when making his supply decision. Carrier k's profit maximization problem is:

$$\underset{\tau^k \geq 0}{\text{maximize}} \quad \sum_{v \in V^k} R_v(\tau)\, \tau_v - \underline{C}^k(\tau^k). \tag{3.1}$$

The well-known first-order conditions for this problem are that marginal revenue equals marginal cost if a positive level of supply is offered, or that

(E-1) if $\tau_v > 0$, then $MR_v = \underline{C}^k_v$

(E-2) if $\underline{C}^k_v > MR_v$, then $\tau_v = 0$,

where

$$MR_v = R_v(\tau) + \sum_{j \in V^k} \tau_j \partial R_j(\tau)/\partial \tau_v, \tag{3.2}$$

$$\underline{C}^k_v = \partial \underline{C}^k(\tau^k)/\partial \tau_v. \tag{3.3}$$

The demand behavior of this market, and, thus, the function $R_v(\tau)$ will be discussed more fully in the next section. Let us focus for the moment on the cost function $\underline{C}^k(\tau^k)$.

Each carrier k produces its vector of outputs τ^k over the subnetwork $H^k(N^k, B^k)$. Let us define

b = an index of arcs, $b \in B$,

e_b = the flow on arc b,

$$e = (e_b \mid b \in B),$$
$$e^k = (e_b \mid b \in B^k),$$

and let us define the following cost measures:

$AC_b(e)$ = the average cost of producing a unit of transportation service on arc $b \in B$, and

$MC_b(e)$ = the marginal total cost of producing this unit of service on arc $b \in B$

$$= \partial\,(\text{Total Cost})/\partial e_b$$

$$= \partial\,(\sum_{i \in B} AC_i(e)e_i + \text{fixed costs})/\partial e_b$$

$$= AC_b(e) + \sum_{i \in B} e_i\ \partial AC_i(e)/\partial e_b. \qquad (3.4)$$

Both $AC_b(e)$ and $MC_b(e)$ are assumed to be continuously differentiable functions. Also, by defining

Q_v = the set of paths between carrier O-D pair v,

$$Q = \bigcup_{v \in V} Q_v,$$

g_q = the flow on carrier path $q \in Q$,

$g = (g_q \mid q \in Q)$, and

$$\lambda_{b,q} = \begin{cases} 1, & \text{if carrier arc } b \in B \text{ is used on path } q \in Q \\ 0, & \text{otherwise,} \end{cases}$$

the following relationships can be stated

$$e_b = \sum_{q \in Q} \lambda_{b,q}\, g_q \qquad \text{for all } b \in B, \qquad (3.5)$$

$$\tau_v = \sum_{q \in Q_v} g_q \qquad \text{for all } v \in V. \qquad (3.6)$$

Finally, let us assume that the costs on a path are the summation of the costs on the arcs which comprise that path, or that

$$AC_q = \sum_{b\varepsilon B} \lambda_{b,q} AC_b(e) \qquad \text{for all } q\varepsilon Q, \text{ and} \qquad (3.7)$$

$$MC_q = \sum_{b\varepsilon B} \lambda_{b,q} MC_b(e) \qquad \text{for all } q\varepsilon Q \qquad (3.8)$$

We will also assume that

(A-4) each carrier attempts to minimize the cost of producing a given vector of outputs,

and thus $\underline{C}^k(\tau^k)$ can be considered the minimum cost function. Thus, we have constructed a two-stage decision process to describe the carriers in which each carrier first decides how to produce τ^k at minimum cost, and then determines the profit-maximizing output level. This description corresponds to the recent conceptual framework proposed by Florian and Gaudry (1983).

The cost function $\underline{C}^k(\tau^k)$ is defined as the solution to the following problem:

$$\underline{C}^k(\tau^k) = \underset{e}{\text{minimum}} \sum_{b\varepsilon B} AC_b(e)e_b + \text{fixed costs}$$

subject to

$$\sum_{q\varepsilon Q_v} g_q = \tau_v \qquad \text{for all } v\varepsilon V^k \ (MC_v^*) \qquad (3.9)$$

$$e_b = \sum_{v\varepsilon V} \sum_{q\varepsilon Q_v} \lambda_{b,q}\, g_q \quad \text{for all } b\varepsilon B^k .$$

Let MC_v^* denote the dual variable of the corresponding path flow conservation constraint in (3.9). The reader will recognize problem (3.9) as the traditional system optimization traffic assignment problem, which has as its first order necessary conditions the Wardropian System Optimality conditions:

(E-3) if $g_q > 0$, $q \varepsilon Q_v$, then $MC_q = MC_v^*$,

(E-4) if $MC_q > MC_v^*$, $q \varepsilon Q_v$, then $g_q = 0$.

That is, the first-order necessary conditions state that all utilized paths between any carrier O-D pair must have their marginal costs equal to the minimum marginal costs for that O-D pair if costs are to be minimized. Therefore, equilibrium conditions (E-1) - (E-4) constitute the first order necessary conditions for each carrier's profit maximization problem, where $\underline{C}_v^k = MC_v^*$ for all $v \varepsilon V$.

If the average cost functions are (strictly) increasing functions of flow (which directly implies that the marginal costs are strictly increasing functions), then it is well-known that problem (3.9) is a (strictly) convex mathematical program and will, thus, yield a unique solution for each vector τ^k. However, if U-shaped average cost functions are used, problem (3.9) will not, in general, have a unique solution. There is the possibility in this case that several flow vectors e will satisfy the first-order conditions (E-3) - (E-4). In this case the function $\underline{C}^k(\tau^k)$ is not uniquely defined, and thus one can only guarantee that $\underline{C}^k$ represents a local minima to (3.9). This lack of uniqueness may raise doubt as to the applicability of this supply-side model to realistic freight markets. We address this issue below.

It is commonly accepted that in the short-run (assumption A-1 of our model), railroads face significant economies of density (average costs decreasing with increasing levels of output). In fact, it is commonly accepted that the average cost functions of most firms exhibit economies of density over some range of output. Thus, U-shaped average cost functions, which imply the existence of a nonconvex total cost function, are typically believed to be the rule and not the exception. In our model, these U-shaped average cost functions imply that multiple local minima will exist for problem (3.9). It will be shown in Section 3.4 that in this situation at least one minimum exists for (3.9) and that algorithms are available to compute a local minimum reliably. However, our ability to define a unique prediction and the function $\underline{C}^k$ as the minimum cost function is effectively lost when dealing with U-shaped average cost functions.

Given that the solution to (3.9) is typically not unique, what can one say about the usefulness of the model of carrier behavior which has been developed in this section? First, a great deal of research has been done in the past few years concerning the issue of characterizing a global minimum solution of nonconvex programs and efficiently solving for this solution. Thus, the framework developed in this section may be amenable to such global solution techniques; future research is needed to ascertain whether recent global solution techniques are practical for solving (3.9). At present, we can only be assured of finding a local minima, and thus our argument that a carrier finds the minimum cost function must be weakened to state that a carrier attempts to find this minimum, as in assumption A-4. This is indeed the weakest aspect of our modeling framework, but appears to be the best one can do given the current state of the art in nonconvex minimization.

However, if the widely held conjecture stated in the previous paragraph that most firms exhibit economies of density in their cost function is true, then virtually all models of economic systems will display the same problem of nonconvex total costs as described above. It is interesting to note that under an assumption of perfect competition, a firm would shut down in the short-run when price falls below average variable costs, and, thus, the short-run marginal cost functions would typically be monotonically increasing, but discontinuous at the shut-down point. This type of marginal cost function would create a convex but non-differentiable total cost function. Therefore, the inclusion of shut-down points would be one way of overcoming the nonconvexity issue. However, most large-scale models must assume that the objective function is continuously differentiable in order to create efficient solution procedures; thus, although the inclusion of these shut-down points is useful to consider in future research, they are not typically analyzed in large-scale economic models due to the computational burden they induce. What is more commonly done is simply to ignore the region of output exhibiting economies of density if this region is small and, thus, to form an approximation of the cost function. The essential point is that the issue of nonconvexities in economic models arises in many circumstances other than problem (3.9) and is a problem which has not been adequately addressed to date. Therefore, although we can conceptually address the issue of nonconvexity in (3.9) by assuming that the transportation market is perfectly competitive, in general we cannot fully address this issue for other market structures and for large-scale problems where discontinuities in the marginal cost functions wreak havoc with solution algorithms. Thus, the best one can do in practice is to state that a local minimum to (3.9) will be found, and argue that such a local minimum is an approximation of the global solution.

In summary, conditions (E-1) - (E-4) will be our description of the carrier's behavior. We next develop a model of the demand-side of the freight market, i.e., of shippers' behavior.

Demand-Side. The shippers are the set of economic agents who decide on the quantity to ship between every pair of regions and on the set of carriers which will move the goods. The shippers must, in general, choose a set of carriers due to the structure of the transportation industry. A single carrier may not be capable of servicing a move from region A to region B, and thus the shipment must be transferred to another carrier to complete the move. Figure 3.2 depicts this situation in which the shipper can choose carrier 1 to move his goods from A to b, and carrier 2 to move the goods from b to their destination B. Thus, the shipper must, in general, choose a sequence of carriers to move his goods. We can conceptualize the shipper decision process by

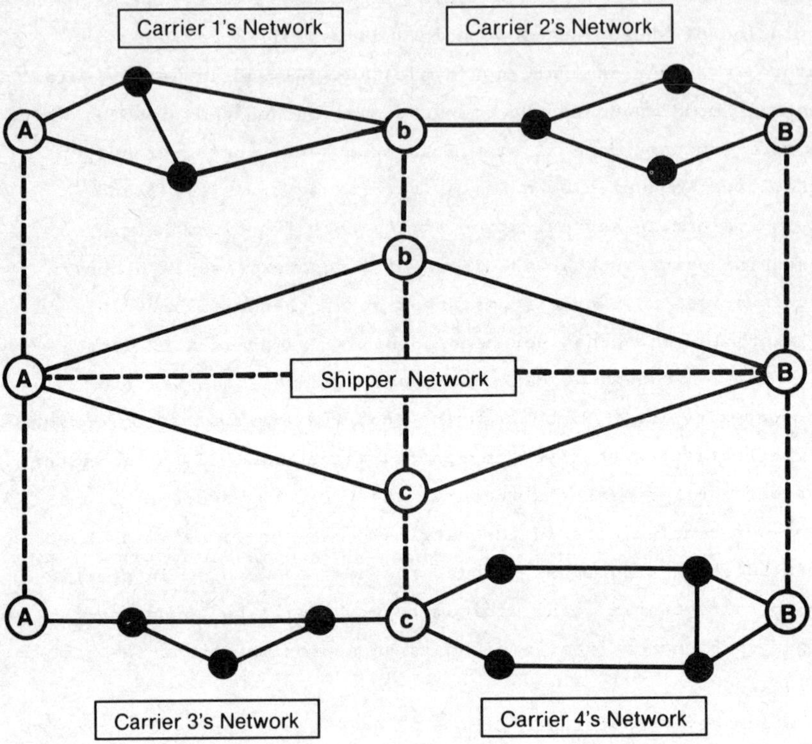

Figure 3.2: Concept of a Shipper Network

thinking of their choice of carriers as a choice of a path on a perceived or shipper network. This network, conceived by Friesz, et al. (1981), consists of the centroids of the regions under study plus those nodes over which the shipper has some control, such as major rail yards, ports, etc. The arcs of this network represent valid O-D moves on the carriers' network. Thus, the demand for service on a carrier O-D pair can be considered as the flow on the shipper arc corresponding to this O-D pair. In Figure 3.2, the flow on arc (A-b) would be considered as the demand for service between carrier 1's O-D pair (1-2). We next formalize the above concepts.

Let L be the set of centroids of the regions under study, and let M be the set of nodes on the shipper network. Thus, $L \subseteq M$. Let A be the set of arcs on the shipper network, and let G(M,A) denote the shipper network. Let us also make the following definitions:

f_a = the flow on shipper arc $a \in A$

f = $(f_a \mid a \in A)$

W = the set of O-D pairs on the shipper network

w = a specific O-D pair, $w \in W$

T_w = the flow between shipper O-D pair $w \in W$

T = $(T_w \mid w \in W)$

P_w = the set of paths between shipper O-D pair $w \in W$

P = $\bigcup_{w \in W} P_w$

p = an index, $p \in P$

h_p = the flow on shipper path $p \in P$

h = $(h_p \mid p \in P)$

$$\delta_{a,p} = \begin{cases} 1, & \text{if the shipper arc } a \in A \text{ is used on path } p \in P \\ 0, & \text{otherwise, and} \end{cases}$$

Δ = $[\delta_{a,p}]$.

Using the above notation, the path O-D pair and arc-path flow constraints can be written as

$$T_w = \sum_{p \in P_w} h_p \qquad \text{for all } w \in W, \text{ and} \tag{3.10}$$

$$f_a = \sum_{p \in P} \delta_{a,p} h_p \qquad \text{for all } a \in A . \tag{3.11}$$

As was stated above, each shipper arc corresponds to a carrier O-D pair. By defining

$$\chi_{a,v} = \begin{cases} 1 \text{ if carrier O-D pair } v \text{ is associated with shipper arc } a \\ 0 \text{ otherwise, and} \end{cases}$$

$$\chi = [\chi_{a,v}] ,$$

where each row and column of χ has only one nonzero entry (a one-to-one correspondence between shipper arcs and carrier O-D pairs), the quantity demanded between any carrier O-D pair is

$$\sum_{a \varepsilon A} \chi_{a,v} f_a = \sum_{a \varepsilon A} \chi_{a,v} \left(\sum_{p \varepsilon P} \delta_{a,p} h_p \right) . \tag{3.12}$$

Also, let us define TC_a to be the total transportation costs on shipper arc a. These costs consist of the dollar charge for transportation service (the freight rate) plus the dollar value of any set of quantifiable level of service attributes. In what follows, the only level of service attribute which will be used is time delay. However, we can, in general, handle any set of well-defined, measurable attributes. By defining

t_v = the time delay on carrier O-D pair $v \varepsilon V$

Φ = the equivalent dollar value of a unit of time delay,

and remembering that R_v denotes the freight rate on carrier O-D pair $v \varepsilon V$, we can write the total cost on shipper arc $a \varepsilon A$ as

$$TC_a = \sum_{v \varepsilon V} \chi_{a,v} (R_v + \Phi t_v). \tag{3.13}$$

Note that each carrier O-D pair $v \varepsilon V$ is defined both by its geographical orientation and its level of service category. Also, each O-D pair is associated with a particular carrier, which in turn is defined as belonging to a particular mode of transportation. Thus, t_v is the time delay which is associated with the mode/carrier/service class represented by O-D pair $v \varepsilon V$.

The easiest way to define t_v would be to assume that it equals some fixed value. However, congestion effects will cause t_v to vary with flow. Thus we would like t_v to fluctuate within the bounds set by the level of service class of which it is a member. Section 3.4 of this chapter will go into detail on how t_v is defined in the operational versions of this model; our purpose here is to introduce the value of the level of service associated with a carrier O-D pair $v \epsilon V$ and allow for the possibility that this service level, in this case time delay, can fluctuate about the mean time delay which characterizes this particular service class.

Finally, the cost TC_p on any shipper path $p \epsilon P$ can be written as

$$\mathbf{TC}_p = \sum_{a \epsilon A} \delta_{a,p} TC_a . \qquad (3.14)$$

Let us assume that

(A-5) the shippers possess no market power in the market for transportation service. That is, the shippers take the price of this service as given in making routing decisions.

Given this assumption, let us consider three possible descriptions of the demand side of the transportation market.

CASE I: Single Carrier Paths

Let us assume that every shipper path between every shipper O-D pair consists of one and only one arc. That is, the shipper must only choose which carrier will service an O-D pair, and does not have to choose a sequence (or path) of carriers. More formally, the matrix Δ has only one nonzero entry in each row and column. Let us also assume that the demand for service between every shipper O-D pair is given, and that a demand function for service on each arc $a \epsilon A$ can be explicitly stated as

$f_a(TC)$ = the demand for service on arc $a \varepsilon A$,

where

$TC = (TC_a | a \varepsilon A)$, and

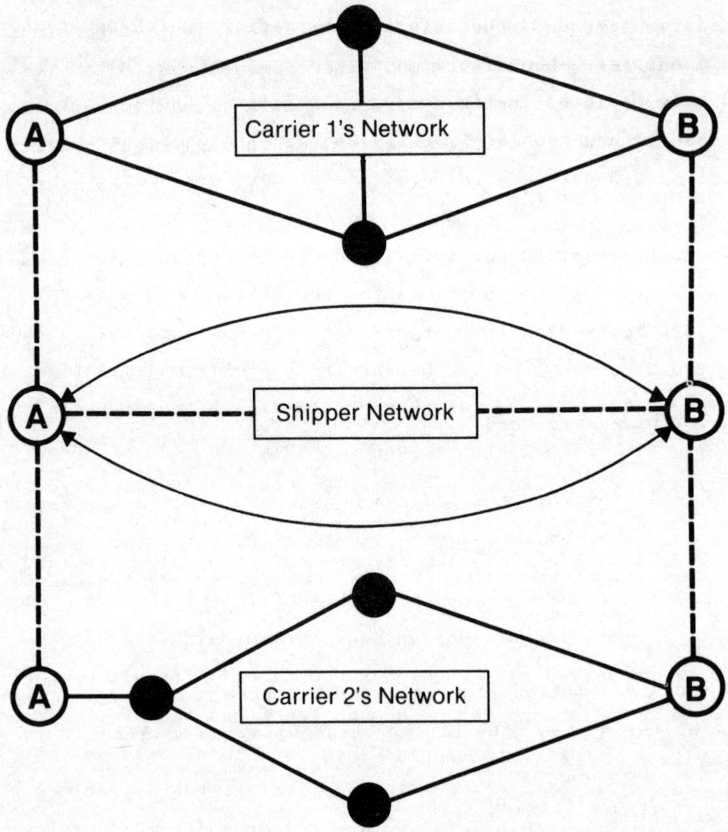

Figure 3.3: Special Case of the Shipper Network

$$T_w = \sum_{p \varepsilon P_w} h_p = \sum_{p \varepsilon P_w} \sum_{a \varepsilon A} \delta_{a,p} f_a(TC) \text{ for all } w \varepsilon W. \quad (3.15)$$

That is, the fixed demand for service between every shipper O-D pair $w \varepsilon W$ is distributed over the set of arcs which connect this O-D pair. Figure 3.3 depicts this situation, which the reader can immediately recognize as a special case of the network structure depicted in Figure 3.2. Therefore, the shipper's decision is how much of T_W it should give to each carrier servicing O-D pair $w \varepsilon W$; no sequence of carriers needs to be created.

Using the relationship (3.12), the functions $f_a(TC)$ can be viewed as the demand functions for the carrier O-D pair $v \varepsilon V$ associated with each shipper arc $a \varepsilon A$. These functions can be inverted and then used in the carrier's profit maximization problem (3.1). Therefore, the construction of this special shipper network corresponds to the assumption that we can define a demand function for each carrier O-D pair. In this case, the problem of predicting freight movements becomes the problem of computing a Nash equilibrium for the set of problems defined by (1). Harker (1984) has shown that an efficient technique exists for the computation of such an equilibrium point.

The above demand-side model has two major difficulties. First, it may be possible to define demand functions for every carrier O-D pair, but the estimation of such functions would be very difficult. Second, for certain commodities, there is often a great deal of switching of carriers and/or modes when moving between a particular shipper O-D pair, and the above approach does not incorporate the decision of a shipper to change carriers. Therefore, although this approach creates a model which is relatively easy to solve, it may not be estimatible or realistic.

CASE II: Multiple Carrier Paths, Elastic Transportation Demand

Let us relax the assumption that each shipper path consists of one and only one arc. In this case, the shipper must choose between the alternative paths in P_w when deciding on the routing of the shipment T_w between O-D pair $w \varepsilon W$. Let us assume

(A-6) the shippers compete noncooperatively for the services of the carriers, and

(A-7) the shippers attempt to minimize the cost of shipping goods between every O-D pair $w \varepsilon W$.

It is well known (see, for example, Devarajan, 1981) that the above two assumptions, along with assumption (A-5), will lead to a flow pattern on the shippers' network which is a Wardropian User-Equilibrium:

(E-5) if $h_p > 0$, then $TC_p = u_w$ for all $w \varepsilon W$, $p \varepsilon P_w$,

(E-6) if $TC_p > u_w$, then $h_p = 0$ for all $w \varepsilon W$, $p \varepsilon P_w$,

where

u_w = the minimum total transportation costs between shipper O-D pair $w \varepsilon W$.

Let us also assume that the demand for movement between each shipper O-D pair $w \varepsilon W$ can be stated as a function of the minimum total costs, or that

$$T_w = T_w(u) \tag{3.16}$$

where

$$u = (u_w | w \varepsilon W).$$

The above demand-side model is the same as the shipper model described in Friesz, Gottfried and Morlok (1981), Friesz, et al (1981), and Gottfried (1983).

The demands for service on a carrier O-D pair are no longer simple functions of TC_a, as in Case I, but are derived through the equilibrium routing of traffic described by conditions (E-5) - (E-6). Thus, we have generalized the demand-side model to allow the shipper to choose a sequence of carriers, but at the same time, we have destroyed the simple model structure described in Case I. The demand functions on a carrier's O-D pair cannot be inverted and the technique described in Harker (1984) cannot be applied to solve (3.1). We will come back to this issue in the next section.

Another problem with this demand-side formulation is that the generation of trips from every region is typically held fixed and separate trip production and attraction constraints are appended for the type of demand function described by (3.16). To include the generation of trips, we will introduce the notion of a spatial price equilibrium into our conceptual framework.

CASE III: Spatial Price Equilibrium

Let us make the following definitions:

L = a set of regions, $\ell \varepsilon L$

π_ℓ = the price of a commodity in region $\ell \varepsilon L$

π = $(\pi_\ell | \ell \varepsilon L)$

$S_\ell(\pi)$ = the supply function in region $\ell \varepsilon L$, and

$D_\ell(\pi)$ = the demand function in region $\ell \varepsilon L$,

and let us assume that

(A-8) the commodity markets in each region L are purely competitive. That is, the shippers take the price of the goods as given when deciding how much of each good to move into or out of a region.

In this case, the shippers act as the equilibrating force in the spatial market, buying and selling goods until a spatial price equilibrium is reached. Using the above notation, the spatial price equilibrium conditions described in (i) and (ii) of Chapter 2 can be written as:

(E-7) if $T_{(i,j)} > 0$, then $\pi_i + u_{(i,j)} = \pi_j$ for all $i,j \in L,(i,j) \in W$,

(E-8) if $\pi_i + u_{(i,j)} > \pi_j$, then $T_{(i,j)} = 0$ for all $i,j \in L,(i,j) \in W$.

Also, the conservation of flow in every region can be written as:

(E-9) $$S_\ell(\pi) - D_\ell(\pi) + \sum_{i \in L} \sum_{(i,\ell) \in W} T_{(i,\ell)} - \sum_{j \in L} \sum_{(\ell,j) \in W} T_{(\ell,j)} = 0 \quad (3.17)$$

Therefore, we have replaced the simple demand function (3.16) by the above behavioral conditions, and in doing so, have incorporated the generation of trips into our model. One can easily see that, of the three cases, this is the most general form of the model. By restricting demand to follow (3.16) or by preventing the shippers from choosing a sequence or path of carriers, we return to the models described in Cases II and I, respectively. We next address the issue of integrating the demand- and supply-side models.

Integration of Supply and Demand. The simplest case of integrating the supplies and demands for transportation is that described in Case I. By simply defining a demand function on each carrier O-D pair, the problem of predicting freight flows and freight rates becomes one of solving a classical Cournot-Nash equilibrium problem. The computations involved in such a problem are not difficult, as Harker (1984) shows, so that this model can be applied to large-scale problems.

However, the assumption that shippers choose the same carriers to initiate and terminate a shipment may not be realistic. Extensive interlinings and transshipments are often made, especially with long-distance shipments. Therefore, we would like to be able to treat the situation described in Case II. Also, we would like to incorporate the producers and consumers of goods into our model so that the generation, distribution, modal split and assignment of trips can be simultaneously predicted. Thus, we will focus on the integration of the supply-side model with the demand-side model described in Case III.

Let us assume that the market for transportation is <u>purely competitive</u>; that is, the carriers take the price as given when solving their profit maximization problem (3.1). This corresponds to the assumption that the (inverse) demand function is inelastic, or that $\partial R_j/\partial \tau_v = 0$ for all $j, v \varepsilon V$. In this case, the carriers will supply up to the point where $MR_v = R_v = \underline{C}_v^k$. Then, by assuming the existence of an 'invisible hand' mechanism and the clearing of all transportation markets, we will have the situation where the freight rates will equal marginal costs at equilibrium.

The assumption of a purely competitive transportation market has allowed us to circumvent the situation in which we must define and estimate a demand function for each carrier O-D pair by <u>decentralizing</u> each agents' decisions. The carrier no longer needs to know the shippers' response to his supply decision in solving (3.1); he has assumed that there will be no response. As will be shown in what follows, this decentralization allows us to create an elegant fixed point representation of the equilibrium conditions. This result is not surprising in that this fixed point representation underlies the study of general competitive market equilibrium (see, for example, Varian, 1978).

If the assumption of pure competition is not adequate, then the carrier must be assumed to have some knowledge of the shippers' response to his supply decision. However, unlike in Case I, the demand functions are not explicitly defined, but are a function of the equilibrium behavior described by conditions (E-5) - (E-9). In this case, it is not yet clear that a simple fixed point problem can be developed or solved for large-scale applications. Future research along the lines of the work done on

two-level mathematical programs by Kolstad and Lasdon (1983) must
be completed before this complex oligopolistic market model can
be successfully used on realistic problems.

Since it is our purpose to create a model which is both
realistic and applicable to large-scale problems, let us introduce the concept of a rate function in order to approximate the
oligopolistic market situation. The rate function, $R_v(\tau)$, is our
à priori assumption concerning the pricing rule of the carriers.
In the case of pure competition, the rate function equals
marginal costs. In the case of fully defined transportation
demands (Case I), the rate function is the inverse demand
function for that O-D pair. Other possible assumptions include
the case where rates are some constant markup over average costs
(i.e., cost-plus pricing), rates being equal to some regulatory
limit, and rates being set equal to some econometrically estimated function of costs, level of service attributes and other
carrier-specific attributes. Chapter 4 will detail the various
ways in which the rate function can be defined. The purpose of
introducing the rate function is to allow one to consider
possible scenarios in which the transportation market is not
assumed to be purely competitive while maintaining the decen-
tralized nature of the market. That is, the carriers still do
not need to know the exact demand function, but are assumed to
price according to the assumed behavior given in the rate
function. As in the purely competitive case, this decentralization of each agent's decisions will allow us to formulate a
single fixed-point representation of our model.

Therefore, we will introduce the concept of a rate function to
approximate the behavior of the freight rates at equilibrium.
In the case of pure competition and Case I, the rate function
will lead to a theoretically correct model of carrier behavior.
In other cases, we are approximating the behavior described by
conditions (E-1) and (E-2). That is, we are assuming that the
carriers, when solving their profit-maximization problem (3.1),

assume that the freight rates will obey the behavior defined by the rate function $R_v(\tau)$. The equilibrum conditions we will use can thus be summarized as follows:

(a) carriers individually minimize costs

$$\text{if } g_q > 0, \text{ then } MC_q = MC_v^* \qquad \text{for all } v \varepsilon V,\ q \varepsilon Q_v$$

$$\text{if } MC_q > MC_v^*, \text{ then } g_q = 0 \qquad \text{for all } v \varepsilon V,\ q \varepsilon Q_v$$

(b) carriers price according to the rate function

$$R_v = R_v(\tau) \qquad \text{for all } v \varepsilon V$$

$$= \underline{C}_v^k = MC_v^* \text{ in the case of pure competition}$$

(c) the transportation markets clear

$$\tau_v = \sum_{a \varepsilon A} \chi_{a,v} f_a \qquad \text{for all } v \varepsilon V \qquad (3.18)$$

(d) shippers individually minimize the cost of shipping goods

$$\text{if } h_p > 0, \text{ then } TC_p = u_w \qquad \text{for all } w \varepsilon W,\ p \varepsilon P_w$$

$$\text{if } TC_p > u_w, \text{ then } h_p = 0 \qquad \text{for all } w \varepsilon W,\ p \varepsilon P_w$$

(e) shippers' behavior results in a spatial price equilibrium

$$\text{if } T_{(i,j)} > 0, \text{ then } \pi_i + u_{(i,j)} = \pi_j \quad \text{for all } i,j \varepsilon L,\ (i,j) \varepsilon W$$

$$\text{if } \pi_i + u_{(i,j)} > \pi_j, \text{ then } T_{(i,j)} = 0 \quad \text{for all } i,j \varepsilon L,\ (i,j) \varepsilon W$$

(f) conservation of flow in every region

$$S_{\ell}(\pi) - D_{\ell}(\pi) + \sum_{i \in L} \sum_{(i,\ell)} T_{(i,j)} - \sum_{j \in L} \sum_{(\ell,j)} T_{(\ell,j)} = 0$$

for all $\ell \in L$.

Given this conceptualization of the freight system, we are now prepared to study questions concerning the equilibrium flows and prices in the economy.

The equilibrium conditions presented in the previous section are a generalization of the usual concept of a spatial price equilibrium by virtue of our inclusion of carriers in the model. Consequently, we shall call the model defined by conditions (a) - (f) the Generalized Spatial Price Equilibrium Model, or GSPEM.

The next section is concerned with alternative mathematical formulations of the equilibrium conditions comprising GSPEM. In creating these mathematical formulations, computational tractability of the model is always kept in mind. Therefore, we will often refer to how certain assumptions lead to more easily solvable mathematical problems. The next subsection presents two procedures for defining the time delay between carrier O-D pairs (t_v) and discusses the implications of these procedures on the computational performance of the model. Then, a nonlinear complementarity formulation of GSPEM is presented which is used to develop existence and uniqueness criteria. A variational inequality formulation is presented next, and it is shown that a special case of this variational inequality leads to an equivalent optimization problem similar to the type encountered in Wardropian user equilibrium and pure spatial price equilibrium. This chapter concludes with a discussion of the relationships between the three mathematical formulations and the various solution methodologies suggested by these formulations.

Definition of Time Delay. The concept of service classes was used to approximate the continuum of service levels which a carrier offers to the shipper on any physical O-D move. Thus, each carrier O-D pair $v \epsilon V$ was defined by its origin, destination and the service class offered. The question which we now address is how to operationalize the definition of the time delay measure (t_v) (our level of service measure) on each carrier O-D pair $v \epsilon V$.

The simplest way to define t_v would be to assume that it equals a fixed value for the entire range of supply (τ_v) offered on O-D pair v. Thus, the shippers' total transportation costs on each arc $a \epsilon A$ could be simply calculated via (3.8) by adding a constant to the rate R_v. However, congestion effects and/or economies of density would most likely cause the delay between carrier O-D pair $v \epsilon V$ to vary with changes in the supply levels. Thus, we would like to allow t_v to be a function of the flows on the carrier network in order to model the fluctuations of t_v about the mean value prescribed for the service class represented by O-D pair $v \epsilon V$.

Let us define

$t_b(e)$ = the time delay on carrier arc $b \epsilon B$, a function of the arc flows e,

and

t_q = the time delay on carrier path $q \epsilon Q$.

We define the path delay measures to be the sum of the arc delays,

$$t_q = \sum_{b \epsilon B} \lambda_{b,q} t_b(e) \qquad \forall\ q \epsilon Q. \qquad (3.19)$$

The first approach which one can use to derive the O-D delay measure t_v is to assume that t_v is derived from the path delay measures t_q. That is, it can be assumed that there exists a mapping Λ_v which takes the path time delay measures t_q and forms the O-D delay measures, or

$$\Lambda_v(t_q | q \varepsilon Q) = t_v. \tag{3.20}$$

This approach to defining t_v will be called endogenous aggregation, since the O-D delay measures are derived from some type of aggregation procedure. One can define many alternative mappings Λ_v. One such mapping is to define t_v as the weighted (by flows) average of the time delays on the paths connecting O-D pair v:

$$t_v = [\sum_{q \varepsilon Q_v} g_q t_q]/\tau_v. \tag{3.21}$$

The second approach to defining time delays is called exogenous aggregation. In this case, the time delays are not derived via some type of aggregation of the carrier path delays as above, but are stated explicitly as functions of the carrier O-D flows

$$t_v = t_v(\tau). \tag{3.22}$$

Since there is a one-to-one correspondence between shipper arc and carrier O-D flows (each row and column of the matrix X has one and only one nonzero entry), (3.22) is equivalent to assuming that the time delay on each shipper arc is a function of the flows on the shipper network, or

$$\tilde{t}_a = \sum_{v \varepsilon V} \chi_{a,v} t_v(\tau) = \sum_{v \varepsilon V} \chi_{a,v} t_v(X^T f) = \tilde{t}_a(f), \tag{3.23}$$

where T denotes the transpose operation, $X^T f = \tau$, and $\tilde{t}_a(f)$ equals the time delay on shipper arc $a \varepsilon A$.

Endogenous aggregation contains more information than does exogenous aggregation in that the change in a carrier arc delay is not reflected in (3.23). Exogenous aggregation has aggregated these carrier arc delays in order to form the simple function $\tilde{t}_a(f)$. Thus, it may seem that endogenous aggregation is always preferable to exogenous aggregation. However, $\tilde{t}_a(f)$ is a much simpler mathematical form than (3.21). We shall illustrate that the use of $\tilde{t}_a(f)$ will allow us to create a simple fixed point representation of GSPEM. Therefore, the choice between exogenous and endogenous aggregation becomes one of choosing between computational tractability and a possibly more realistic representation of time delays.

3.2. Nonlinear Complementarity Formulation

Having introduced all relevant notation and defined all necessary functions, we now turn to the issue of developing a compact mathematical formulation of GSPEM. This section places GSPEM into the form of a complementarity problem; let us begin by defining this concept.

A complementarity problem (CP) is the problem of finding a vector of variables $x = (x_1, x_2, \ldots, x_n)^T$ such that for the vector function $F(x) = (F_1(x), F_2(x), \ldots, F_n(x))^T$ the following conditions hold, where T denotes the transpose operation:

$$\begin{aligned} F(x)^T \cdot x &= 0 \\ F(x) &\geq 0 \\ x &\geq 0. \end{aligned} \tag{3.24}$$

If the function F is linear, then (3.24) is called a linear complementarity problem (LCP); if it is nonlinear, then (3.25) is a nonlinear complementarity problem (NCP).

The following theorem shows that the equilibrium conditions (a) - (f) can be put into the form of a CP.

Theorem 3.1

If it is assumed that

(i) TC_a is strictly positive $\forall\ a \varepsilon A$,
(ii) MC_b is strictly positive $\forall\ b \varepsilon B$, and

iii) <u>supply never exceeds demand for free goods:</u>
$\pi_\ell = 0 \rightarrow D_\ell(\pi) \geq S_\ell(\pi) \quad \forall\ \ell\varepsilon L,$

<u>then the equilibrium conditions (a) - (f) plus (3) - (9) are equivalent to the following CP:</u>

$$F^1_\ell = S_\ell(\pi) - D_\ell(\pi) + I_\ell - E_\ell = 0 \quad \forall\ \ell\varepsilon L \tag{3.25}$$

$$F^2_p = \pi_i + TC_p - \pi_j \qquad \forall\ i,j\varepsilon L,\ (i,j)\varepsilon W,\ p\varepsilon P_{(i,j)} \tag{3.26}$$

$$F^3_v = \sum_{q\varepsilon Q_v} g_q - \sum_{a\varepsilon A} \chi_{a,v} \left(\sum_{p\varepsilon P} \delta_{a,p} h_p \right) \quad \forall\ v\varepsilon V, \tag{3.27}$$

$$F^4_q = MC_q - MC^*_v \qquad \forall\ v\varepsilon V,\ q\varepsilon Q_v, \tag{3.28}$$

$$F^T = \lfloor (F^1_\ell | \ell\varepsilon L);\ (F^2_p | p\varepsilon P);\ (F^3_v | v\varepsilon V);\ (F^4_q | q\varepsilon Q) \rfloor,$$

$$x^T = \lfloor \pi;\ h;\ MC^*;\ g \rfloor,$$

$$F^T \cdot x = \sum_{\ell\varepsilon L} F^1_\ell \pi_\ell + \sum_{i,j\varepsilon L} \sum_{(i,j)\varepsilon W} \sum_{p\varepsilon P_{(i,j)}} F^2_p h_p$$

$$+ \sum_{v\varepsilon V} F^3_v MC^*_v + \sum_{v\varepsilon V} \sum_{q\varepsilon Q_v} F^4_q g_q = 0. \tag{3.29}$$

$$F \geq 0$$

$$x \geq 0,$$

where $MC^* = (MC_v^* | v \varepsilon V)^T$,

$$I_\ell = \text{imports into region } \ell \varepsilon L$$

$$= \sum_{i \varepsilon L} \sum_{(i,\ell) \varepsilon W} \sum_{p \varepsilon P_{(i,\ell)}} h_p,$$

and $$E_\ell = \text{exports from region } \ell \varepsilon L$$

$$= \sum_{j \varepsilon L} \sum_{(\ell,j) \varepsilon W} \sum_{p \varepsilon P_{(\ell,j)}} h_p .$$

Proof

It is trivial to show that the last summation in (3.29) yields equilibrium condition (a). This summation, along with the nonnegativity of F and x implies $(MC_q - MC_v^*) g_q = 0$, $MC_q - MC_v^* \geq 0$, $g_q \geq 0$ for all $v \varepsilon V$, $q \varepsilon Q_v$. If $g_q > 0$, then $MC_q = MC_v^*$; and if $MC_q > MC_v^*$, then $g_q = 0$.

Condition (b) follows directly from the definition of the rate for each carrier O-D pair.

It is also trivial to show that the second summation in (3.29) yields conditions (d) and (e). This summation and the nonnegativity of F and x implies $(\pi_i + TC_p - \pi_j) h_p = 0$, $\pi_i + TC_p - \pi_j \geq 0$, $h_p \geq 0$ for all $i,j \varepsilon L$, $(i,j) \varepsilon W$, $p \varepsilon P_{(i,j)}$. Thus, if $h_p > 0$, then $TC_p = \pi_j - \pi_i$ and if $TC_p > \pi_j - \pi_i$, then $h_p = 0$. Also, if $h_p > 0$, this implies by (3.10) that $T_{(i,j)} > 0$, and if $T_{(i,j)} = 0$, then $h_p = 0$ for all $p \varepsilon P_{(i,j)}$. Taken together, these relationships imply that

$$h_p > 0 \Rightarrow T_{(i,j)} > 0$$

$$\Rightarrow TC_p = u_{(i,j)} = \pi_j - \pi_i$$

and

$$u_{(i,j)} > \pi_j - \pi_i \Rightarrow TC_p > \pi_j - \pi_i \Rightarrow h_p = 0$$

$$\forall\ p\varepsilon P_{(i,j)} \Rightarrow T_{(i,j)} = 0$$

which is just a restatement of conditions (d) and (e).

The first function defined in F is allowed to be greater than zero, but condition (f) says this function must hold with equality. Assume the contrary:

$$S_\ell(\pi) - D_\ell(\pi) + I_\ell - E_\ell > 0. \tag{3.30}$$

Since $h \geq 0$, both I_ℓ and E_ℓ are nonnegative. For (3.30) to hold, $\pi_\ell = 0$. This implies by assumption (iii) that $D_\ell(\pi) - S_\ell(\pi) \geq 0$. Therefore, rewriting (21), we have

$$I_\ell > D_\ell(\pi) - S_\ell(\pi) + E_\ell \geq 0; \tag{3.31}$$

that is, region ℓ must be an importer. For this to be true, at least one path in $p\varepsilon P_{(i,\ell)}$ for any $i\varepsilon L$, $(i,\ell)\varepsilon W$ must have positive flow. This implies, by the second relation in the CP that

$$\pi_i + TC_p = \pi_\ell = 0 \tag{3.32}$$

However, assumption (i) states that $TC_a > 0\ \forall\ a\varepsilon A$ which implies $TC_p > 0$. This fact, along with $\pi_i \geq 0$, implies (3.32) cannot hold, thus contradicting (3.31). Therefore, the first function defined in F must always hold with equality, which is just our flow conservation contraint.

The third function defined in F is the conservation of flow between the shipper and carrier networks. This function is simply a combination of condition (c) and (3.6) and (3.11). This function must also hold with equality. Assume the contrary, then it must be the case that

$$\sum_{q\varepsilon Q_v} g_q > \sum_{a\varepsilon A} \chi_{a,v} \Big(\sum_{p\varepsilon P} \delta_{a,p}\, h_p \Big) \tag{3.33}$$

and $MC_v^* = 0$. Since $h \geq 0$, there must exist at least one $q \varepsilon Q_v$ with $g_q > 0$. This implies by the last relation in $F^T \cdot x$ that

$$MC_q = MC_v^* = 0. \tag{3.34}$$

But by assumption (ii), $MC_a > 0 \Rightarrow MC_q > 0$ which contradicts (3.34). Therefore, (3.33) cannot hold and the third function defined in F must hold with equality, yielding condition (c).

Q.E.D.

Condition (iii) was first used by Friesz, Tobin, Smith and Harker (1983) to prove that the standard spatial price equilibrium problem can be cast into the form of a NCP. Let us now turn to the question of the existence and uniqueness of a solution to GSPEM.

3.3 Existence and Uniqueness of an Equilibrium

In this section, the conditions under which an equilibrium to GSPEM is assured, and the conditions which will assure a unique equilibrium, are developed. Let us start with the existence question.

In a recent paper, Smith (1984) has developed some general conditions under which a solution to a complementarity problem is guaranteed. Thus, let us start our discussion of existence by describing Smith's results.

Smith defines the complementarity problem (CP) by defining $F:R_+^n \rightarrow R_n$ and $x \varepsilon R_+^n$, and saying that x is a CP - solution for F if and only if (i) $F(x) \varepsilon R_+^n$, and (ii) $F(x)^T \cdot x = 0$, where

R = the real numbers,

R^n = n-dimensional Euclidean space,

R^n_+ = the nonnegative orthant of n-dimensional Euclidian space, and

n = a positive integer

The function F belongs to the class of continuous functions mapping R^n_+ into R_n, this class being denoted by C^n_+.

Letting m be a positive integer, Smith (1984) defines the concept of an exceptional sequence (ES), the definition being:

For any F: $R^n_+ \rightarrow R^n$, a sequence $\{x^m\}$ on R^n_+ with $|x^m| = m$ for all [positive integers] m is said to be an exceptional sequence for F if each $x^m = (x^m_i)$ satisfies the following two conditions for some positive scalar $\lambda_m > 0$,

$$x^m_i > 0 \Rightarrow F_i(x^m) = -\lambda_m x^m_i \tag{3.35}$$

$$x^m_i = 0 \Rightarrow F_i(x^m) \geq 0 \ . \tag{3.36}$$

He also defines the ES-condition by saying that a function F: $R^n_+ \rightarrow R^n$ satisfies this condition if there exists no exceptional sequence for F.

The two main results of Smith's (1984) paper are his Theorem 4.3 and Corollary 4.4. We will state his results without proof in the following two lemmas:

Lemma 3.1 For any F ε C^n_+, either F has a CP-solution or there exists an exceptional sequence for F.

Lemma 3.2 Each F ε C^n_+ satisfying the ES-condition has a CP-solution.

The following result will prove useful in establishing the existence of a solution to GSPEM.

Lemma 3.3 If the sequence $\{x^m\}$ is bounded ($\|x^m\| < \infty$), then this sequence cannot be an exceptional sequence.

Proof

By definition, an exceptional sequence is one for which $\|x^m\| = m$ for all positive integers m. Let $m \to \infty$. Then, to be an exceptional sequence, it must be the case that $\|x^m\| \to \infty$. However, it is assumed that $\|x^m\| < \infty$, and thus a contradiction is reached.

Q.E.D.

Using the above lemmas we now prove the following result:

Theorem 3.2

If

(i) TC_a is positive, continuous and bounded away from zero $\forall$ $a \varepsilon A$,

(ii) $\pi_\ell > \bar{\pi}_\ell \to S_\ell(\pi) \geq D_\ell(\pi)$ $\forall$ $\ell \varepsilon L$; that is, there is some price in each subregion whereby any price above this value implies the supply in this region is at least as great as demand,

(iii) $S_\ell(\pi)$ and $D_\ell(\pi)$ are continuous $\forall$ $\ell \varepsilon L$, and

(iv) $MC_b(e)$ is positive, continuous and bounded away from zero $\forall$ $b \varepsilon B$,

then a solution to GSPEM exists.

Proof

To show that a solution to GSPEM exists, it is sufficient to show that a CP-solution exists for the CP defined in Theorem 3.1. To show that this CP-solution exists, it suffices to show, by Lemma 3.2, that F satisfies the ES-condition.

Let us assume the contrary. Then there must exist an exceptional sequence $\{(x^m)\} = \{\pi^m; h^m; MC^{*m}; g^m\}$ for F. Looking first at the price sequence $\{\pi^m\}$, we want to show that this sequence is bounded. For if $\pi_\ell^m \to \infty$ holds for any $\ell \varepsilon L$, then choosing any $\ell(m) \varepsilon L$ with $\pi^m_{\ell(m)} = \max \{\pi^m_\ell | \ell \varepsilon L\}$, it follows that $\pi^m_{\ell(m)} \to \infty$ and hence that $\pi^m_{\ell(m)} > \bar{\pi}^m_{\ell(m)} \geq 0$ for all m sufficiently large. This in turn implies from assumption (ii) that $S_{\ell(m)}(\pi^m) - D_{\ell(m)}(\pi^m) \geq 0$.

For an ES to occur, condition (3.35) must hold for $\pi^m_{\ell(m)}$, or

$$S_{\ell(m)}(\pi^m) - D_{\ell(m)}(\pi^m) + I^m_{\ell(m)} - E^m_{\ell(m)} = -\lambda_m \pi^m_\ell < 0 \qquad (3.37)$$

where $I^m_{\ell(m)}$ and $E^m_{\ell(m)}$ are the imports and exports as defined in Theorem 1. The nonnegativity of $I^m_{\ell(m)}$ with the condition that $S_{\ell(m)}(\pi^m) - D_{\ell(m)}(\pi^m) \geq 0$ implies that $E^m_{\ell(m)} > 0$. However, for $E^m_{\ell(m)} > 0$, it must be true that at least one $p \varepsilon P(\ell(m),j), j \varepsilon L, (\ell(m),j) \varepsilon W$ has $h_p > 0$. From condition (3.35) this implies that for this path $\pi^m_{\ell(m)} + TC_p - \pi^m_j < 0 \Rightarrow TY^m_j > \pi^m_{\ell(m)}$ due to the positivity of TC_p implied by assumption (i). This condition contradicts the maximality of $\pi^m_{\ell(m)}$, and thus $E^m_{\ell(m)} > 0$ is not possible, and we obtain a contradiction. Therefore, the price sequence $\{\pi^m\}$ must be bounded.

Next, we want to show that the sequence $\{h^m\}$ is bounded. Assume the contrary. Then there must exist at least one $p \varepsilon P$ such that $h^m_p \to \infty$. From condition (3.35), this implies that for this path $p \varepsilon P(i,j)$

$$\pi^m_i + TC_p - \pi^m_j = -\lambda_m h^m_p . \qquad (3.38)$$

Condition (3.38) implies that

$$\pi_i^m + TC_p - \pi_j^m < 0$$

since $\lambda_m h_p^m$ is positive, or that

$$\pi_j^m > \pi_i^m + TC_p > 0 \ . \tag{3.39}$$

Therefore, the price at the destination node is positive. This implies via the first summation in (3.29) and condition (3.35) that

$$S_j(\pi^m) - D_j(\pi^m) + I_j^m - E_j^m = -\lambda_m \pi_j^m \ . \tag{3.40}$$

Since $h_p^m \to \infty$ for $p \varepsilon P_{(i,j)}$, this implies that $I_j^m \to \infty$. Since $\{\pi^m\}$ is bounded, $S_j(\pi^m) - D_j(\pi^m)$ is bounded, and thus the only way for (3.40) to hold as $I_j^m \to \infty$ is for $E_j^m \to \infty$. This implies that there must exist at least one path $p \varepsilon P_{(j,\ell)}$ for some $\ell \varepsilon L$ such that the flow on this path is tending towards infinity. Also, condition (3.39) implies that $\pi_\ell^m > \pi_j^m$. By induction, the above argument will create a sequence of nodes i, j, $\ell(j+1)$, $\ell(j+2)$, ..., $\ell(j+k)$, ... such that $\pi^m_{\ell(j+k+1)} > \pi^m_{\ell(j+k)}$. This sequence terminates at $\ell(m) \varepsilon L$ such that $\pi^m_{\ell(m)} = \max \{\pi_\ell^m | \ell \varepsilon L\}$. At this node, condition (3.35) implies that

$$S_{\ell(m)}(\pi^m) - D_{\ell(m)}(\pi^m) + I^m_{\ell(m)} - E^m_{\ell(m)} = -\lambda_m \pi^m_{\ell(m)} . \tag{3.41}$$

Since $I^m_{\ell(m)} \to \infty$, it must be the case that $E^m_{\ell(m)} \to \infty$ for (3.41) to hold. However, this implies by (3.39) that there exists a destination node such that the price at this node is greater than $\pi^m_{\ell(m)}$. This requirement constitutes a contradiction of the maximality of $\pi^m_{\ell(m)}$, and thus $E^m_{\ell(m)} \to \infty$ is not possible, and we obtain a contradiction. Therefore, the shipper path flow sequence $\{h^m\}$ must be bounded.

Next, we want to show that the sequence $\{g^m\}$ is bounded. Assume the contrary. Then there must exist at least one $q \varepsilon Q$ such that the sequence $g_q^m \to \infty$. The third summation in (3.29) and conditions (3.35) and (3.36) imply

$$MC_v^{*m} > 0 \Rightarrow \sum_{q \varepsilon Q_v} g_q^m - H_v^m = -\lambda_m MC_v^{*m} \tag{3.42}$$

and

$$MC_v^{*m} = 0 \Rightarrow \sum_{q \varepsilon Q_v} g_p^m - H_v^m \geq 0, \tag{3.43}$$

where

$$H_v^m = \sum_{a \varepsilon A} \chi_{a,v} \sum_{p \varepsilon P} \delta_{a,p} h_p^m .$$

Since $\{h_p^m\}$ is bounded, H_v^m is bounded. Thus, as $g_q^m \to \infty$, the left-hand sides of (3.42) and (3.43) tend to positive infinity. Clearly, (3.42) cannot hold, and thus $MC_v^{*m} = 0$. The last summation in (3.29) and condition (3.35) imply

$$MC_q^m - MC_v^{*m} = MC_q^m - 0 = MC_q^m = -\lambda_m g_q^m . \tag{3.44}$$

Assumption (iv) implies that MC_q^m is positive and bounded away from zero. However, the right-hand side of (3.44) can never be positive as $g_q^m \to \infty$, and a contradiction is reached. Therefore, the carrier path flow sequence $\{g^m\}$ must be bounded.

Finally, we want to show that the sequence $\{MC^{*m}\}$ is bounded. Assume the contrary. Then there must exist at least one $v \varepsilon V$ with $MC_v^{*m} \to \infty$. Condition (3.42), along with the boundedness of H_v^m and $\{g^m\}$, implies that $\lambda_m \to 0$. The last summation in (3.29) implies for all $q \varepsilon Q_v$

$$g_q^m > 0 \Rightarrow MC_q^m - MC_v^{*m} = -\lambda_m g_q^m \tag{3.45}$$

and

$$g_p^m = 0 \Rightarrow MC_q^m - MC_v^{*m} \geq 0 . \tag{3.46}$$

The boundedness of $\{g^m\}$ and assumption (iv) implies MC_q^m is bounded. Clearly, if $g_q^m = 0$ for some $q\epsilon Q_v$, then as $MC_v^{*m} \to \infty$, (3.46) cannot be met and a contradiction is reached. If $g_q^m > 0$, (3.45) and the boundedness of $\{g^m\}$ imply that $\lambda_m \to \infty$ as $MC_v^{*m} \to \infty$. However, this contradicts the above condition that $\lambda_m \to 0$. Therefore, a contradiction is reached and $\{MC_v^{*m}\}$ must be bounded.

Thus, we have established that if $\{x^m\} = \{\pi^m; h^m; MC^{*m}; g^m\}$ is an exceptional sequence, it must be a bounded sequence. However, Lemma 3.3 implies that this condition is a contradiction, and thus no exceptional sequences exist. By Lemma 3.2, this implies that there exists a solution to the CP defined in Theorem 3.1.

Q.E.D.

Therefore, if costs are strictly positive and supply eventually exceeds demand at every region, then the existence of a solution is assured. It should be noted that other than the conditions on the extreme ends of the supply and demand functions, no mention has been made about the shape of the functional forms used. That is, convexity, monotonicity, etc., are not required to insure existence. Existence criteria could also be developed from the variational inequality formulation of the next section, but these criteria would prove to be more restrictive than those given here and, hence, are not reported.

Let us now turn to the question of the uniqueness of a solution to GSPEM.

To begin our discussion of the uniqueness of an equilibrium for GSPEM, let us make the following definitions:

(i)   A function $F: R^m \to R^n$ is said to be monotone (increasing) if, for all vectors $x\epsilon R^n$, $y\epsilon R^n$, in a feasible set Ω, the following condition holds

$$[F(x) - F(y)]^T \cdot (x-y) \geq 0 . \qquad (3.47)$$

(ii) A monotone function F, as defined above, is said to be strictly monotone if

$$[F(x) - F(y)]^T \cdot (x-y) \geq 0 .$$

implies that x = y.

(iii) A variational inequality problem (VIP) is to find a vector $x \varepsilon \Omega$ such that

$$F(x)^T \cdot (y-x) \geq 0 . \quad (3.48)$$

holds for all $y \varepsilon \Omega$.

Karamardian (1971) has shown that every complementarity problem (CP) is a VIP.

A well-known result in variational inequality theory concerning the question of the uniqueness of a solution is (see, for example, p. 14 of Kinderlehrer and Stampacchia, 1980):

Lemma 3.4 If F is a strictly monotone function, then the variational inequality problem defined by (3.48) has at most one solution.

We now state our main uniqueness result.

Theorem 3.3

If

(i) $(S_\ell(\pi) - D_\ell(\pi))$ is a strictly monotone function $\forall\ \ell \varepsilon L$,

(ii) $MC_b(e)$ is a strictly monotone function $\forall\ b \varepsilon B$, and

(iii) $(R_v + \phi\, t_v - MC_v^*)$ <u>is a strictly monotone function</u> $\forall$ $v \epsilon V$,

<u>then there exists at most one solution to GSPEM.</u>

<u>Proof</u>

Since every CP is a VIP, let us calculate (3.47) for the function F defined in Theorem 3.1. Letting $x = (\bar{\pi};\ \bar{h};\ \overline{MC}^*;\ \bar{g})$ and $y = (\pi;\ h;\ MC^*;\ g)$, we have

$$[F(x) - F(y)]^T \cdot (x-y) = \sum_{\ell \epsilon L} [F_\ell^1(x) - F_\ell^1(y)]\ (\bar{\pi}_\ell - \pi_\ell)$$

$$+ \sum_{p \epsilon P} [F_p^2(x) - F_p^2(y)]\ (\bar{h}_p - h_p) + \sum_{v \epsilon V} [F_v^3(x) - F_v^3(y)]\ (\overline{MC}_v^* - MC_v^*)$$

$$+ \sum_{q \epsilon Q} [F_q^4(x) - F_q^4(y)]\ (\bar{g}_q - g_q)\ ,$$

where F_ℓ^1, F_p^2, F_v^3, and F_q^4 are defined in Theorem 1.

Combining terms, we have

$$[F(x) - F(y)]^T \cdot (x-y) = \sum_{\ell \epsilon L} [(\bar{S}_\ell(\pi) - \bar{D}_\ell(\pi)) - (S_\ell(\pi) - D_\ell(\pi))]\ (\bar{\pi}_\ell - \pi_\ell)$$

$$+ \sum_{p \epsilon P} [(\overline{TC}_p - \overline{\underline{MC}}_p) - (\underline{TC}_p - \underline{MC}_p)]\ (\bar{h}_p - h_p)$$

$$+ \sum_{q \epsilon Q} [\overline{MC}_q - MC_q]\ (\bar{g}_q - g_q)\ , \qquad (3.49)$$

where TC_p is defined by (3.14) and

$$\underline{MC}_p = \sum_{a \epsilon A} \delta_{a,p} \sum_{v \epsilon V} \chi_{a,v}\, MC_v^*\ . \qquad (3.50)$$

= the minimum marginal transportation costs on shipper path $p \epsilon P$.

Using the relationships defined earlier, we can rewrite (3.49) as

$$[F(x) - F(y)]^T \cdot (x-y) = \sum_{\ell \varepsilon L} [(S_\ell(\bar{\pi})-D_\ell(\bar{\pi})) - (S_\ell(\pi)-D_\ell(\pi))] (\bar{\pi}_\ell - \pi_\ell)$$

$$+ \sum_{v \varepsilon V} [(\bar{R}_v + \Phi\bar{t}_v - \overline{MC}^*_v) - (R_v + \Phi t_v - MC^*_v)] (\bar{\tau}_v - \tau_v)$$

$$+ \sum_{b \varepsilon B} [MC_b(\bar{e}) - MC_b(e)] (\bar{e}_b - e_b) . \qquad (3.51)$$

By assumptions (i) - (iii), each term in (3.51) is strictly monotone, and thus F must be strictly monotone. By Lemma 3.4, we know that F being strictly monotone implies a unique solution. Thus, the solution to GSPEM is unique if it exists.

Q.E.D.

The following corollary states some sufficient, but not necessary, conditions to insure uniqueness.

Corollary 3.1

If assumptions (i)and (ii) of Theorem 3 hold, plus either

(iii) $(R_v-MC^*_v)$ is a monotone function $\forall$ $v \varepsilon V$, and

(iv) ϕt_v is a strictly monotone function $\forall$ $v \varepsilon V$,

or

(v) $(R_v-MC^*_v)$ is a strictly monotone function $\forall$ $v \varepsilon V$,

and

(vi) ϕt_v is a monotone function $\forall$ $v \varepsilon V$,

then the conclusion of Theorem 3 holds.

Proof

Both (iii) - (iv) and (v) - (vi) imply that $(R_v + \Phi t_v - MC_v^*)$ is a strictly monotone function $\forall$ $v \epsilon V$, and thus all the sufficient conditions for Theorem 3 are met.

Q.E.D.

Therefore, if the rate charged for each O-D move is at least as great as the marginal cost for that move and there is a strictly increasing value of time delay along with assumptions (i) and (ii) of Theorem 3.3, then uniqueness is assured. This set of conditions allows for marginal costs pricing, $R_v - MC_v^* = 0$, since this is a monotone (nondecreasing) function. If the time delay functions are not strictly increasing, then conditions (v) - (vi) state that the rate must be greater than the marginal costs, this difference being a strictly increasing function. Again, these are not necessary conditions, but they are sufficient to guarantee uniqueness.

The major problem in insuring the uniqueness of an equilibrium for GSPEM lies in assumption (ii). In many freight industries, notably railroads, there are often U-shaped average cost curves on arcs. That is, there are economies of density in freight movements over at least some range of output. This fact implies the marginal cost function is also U-shaped, and thus not strictly monotone. Therefore, if the industries under study have non-monotone functions, uniqueness cannot be assured. However, the existence of a solution does not depend on monotonicity. Existence can be assured with non-monotonic functions, as long as the conditions of Theorem 2 are met.

3.4 Variational Inequality Formulation

In this section, an alternative formulation of the Generalized Spatial Price Equilibrium Model will be developed. This alternative formulation is that of a variational inequality problem which is to find a vector $x \epsilon \Omega$, Ω being a feasible set, such that the following condition holds for all vectors $y \epsilon \Omega$:

$$F(x)^T \cdot (y-x) \geq 0 \ . \tag{3.52}$$

As was pointed out in the previous section, Karamardian (1971) has shown that every complementarity problem (CP) is a variational inequality problem (VIP). However, the converse of this statement is not true in general. That is, it is not generally true that every VIP can be stated as a CP. Therefore, this section will develop a VIP formulation of GSPEM which is not the formulation which derives from the CP of Section 3.3.

The formulation developed in this section is important for two reasons. First, as will be discussed in Chapter 5, the VIP formulation of GSPEM yields, under certain assumptions, an optimization-based solution algorithm which is capable of solving large-scale problems. Second, the VIP formulation is closely related to the type of extremal formulation of the spatial price equilibrium model discussed by Samuelson (1952).

Let us assume that the supply and demand functions in each region $\ell \epsilon L$ are <u>invertible</u> and let us define

$$\begin{aligned}
S &= (\ldots, S_\ell, \ldots)^T, \\
D &= (\ldots, D_\ell, \ldots)^T, \\
\Psi_\ell(S) &= \text{the inverse supply function for region } \ell, \\
\Psi &= (\ldots, \Psi_\ell(S), \ldots)^T, \\
\Theta_\ell(D) &= \text{the inverse demand function for region } \ell, \\
\Theta &= (\ldots, \Theta_\ell(d), \ldots)^T,
\end{aligned}$$

and let us not treat MC_v^* as a decision variable as in the CP formulation, but let us define MC_v^* to be a function:

$$\begin{aligned}
MC_v^* &= \min_{q \epsilon Q_v} \{MC_q\} \\
&= \text{the minimum marginal cost between carrier O-D pair v.}
\end{aligned}$$

The shipper equilibrium conditions (d) - (f) in Section 3.1 can be rewritten as (see, for example, Florian and Los, 1982):

$$(g)\quad S_\ell - D_\ell + I_\ell - E_\ell = 0 \qquad \forall\ \ell\epsilon L \tag{3.53}$$

$$(h)\quad \text{if } h_p > 0, \text{ then } \Psi_i(S) + TC_p = \Theta_j(D)$$

$$\forall\ i,j\epsilon L,\ (i,j)\epsilon W,\ p\epsilon P_{(i,j)}$$

$$\text{if } \Psi_i(S) + TC_p > \Theta_j(D), \text{ then } h_p = 0$$

$$\forall\ i,j\epsilon L,\ (i,j)\epsilon W,\ p\epsilon P_{(i,j)}\ .$$

Thus, $\Psi_i(S)$ can be interpreted as the price the shipper pays for a commodity at the origin (the supply price) and $\Theta_j(D)$ is the price the shipper receives when the commodity is sold at the destination (the demand price).

The vector of decision variables we are considering is thus $x = (S;\ D;\ f;\ e)^T$ and this vector must lie in the feasible set Ω, which is comprised of the flow conservation constraints (3.8), (3.10), (3.18) and (3.53) plus the definitional constraints (3.5) and (3.11); that is

$$\Omega = \{x \mid D_\ell - S_\ell + \sum_{j\epsilon L}\ \sum_{(\ell,j)\epsilon W} T_{(\ell,j)} - \sum_{i\epsilon L}\ \sum_{(i,\ell)\epsilon W} T_{(i,\ell)} = 0 \qquad \forall\ \ell\epsilon L;$$

$$T_w = \sum_{p\epsilon P_w} h_p \qquad \forall\ w\epsilon W;$$

$$f_a = \sum_{p\epsilon P} \delta_{a,p}\, h_p \qquad \forall\ a\epsilon A;$$

$$\tau_v = \sum_{a\epsilon A} \chi_{a,v}\, f_a \qquad \forall\ v\epsilon V;$$

$$\tau_v = \sum_{q\epsilon Q_v} g_q \qquad \forall\ v\epsilon V.$$

$$e_b = \sum_{q\epsilon Q} \lambda_{b,q}\, g_q \qquad \forall\ b\epsilon B;$$

$$x \geq 0\}\ .$$

Many researchers, such as Florian and Los (1982), consider a slightly different specification of condition (g) in analyzing spatial price equilibrium. By defining path $p(\ell)$ which has both its origin and destination at the same node ℓ, the flow conservation equation (g) is written as:

$$S_\ell = E_\ell + h_{p(\ell)} \ . \tag{3.54}$$

$$D_\ell = I_\ell + h_{p(\ell)} \ . \tag{3.55}$$

Condition (3.54) says that the supply at a node equals the flow being exported to other nodes plus the flow which leaves node ℓ and returns to this same node. Likewise, the demand to a node equals the flow into this node from all other nodes, including itself. The following lemma shows that (3.54) - (3.55) and (3.53) are equivalent representations of the flow conservation at node ℓ.

<u>Lemma 3.5</u> <u>Equations</u> (3.53) <u>and</u> (3.54) - (3.55) <u>are completely equivalent</u>.

<u>Proof</u>

To show that (3.54) - (3.55) implies (3.53), let us rewrite (3.54) and (3.55) as

$$h_{p(\ell)} = S_\ell - E_\ell \tag{3.56}$$

$$h_{p(\ell)} = D_\ell - I_\ell \tag{3.57}$$

Subtracting (3.57) from (3.56), we have

$$h_{p(\ell)} - h_{p(\ell)} = 0 = S_\ell - D_\ell + I_\ell - E_\ell$$

which is equation (3.53).

To show that (3.53) implies (3.54) - (3.55), let us define

$$h^1_{p(\ell)} = S_\ell - E_\ell \tag{3.58}$$

$$h^2_{p(\ell)} = D_\ell - I_\ell \tag{3.59}$$

Equation (3.53) can be rewritten as

$$(S_\ell - E_\ell) - (D_\ell - I_\ell) = 0$$

or

$$h^1_{p(\ell)} - h^2_{p(\ell)} = 0 ,$$

which implies $h^1_{p(\ell)} = h^2_{p(\ell)} = h_{p(\ell)}$, and thus (3.53) is equivalent to (3.54) - (3.55).

Q.E.D.

Therefore, we can either treat condition (g) as one equation or two. This fact will be important in formulating GSPEM as a VIP.

Let us assume that the costs on the path $p(\ell)$ are zero, or that

$$TC_{p(\ell)} = 0 \quad \forall \; \ell \varepsilon L.$$

This assumption says that there are costless transactions within a region, which is also the assumption underlying the original formulation of shippers' behavior given by conditions (d) - (f). Therefore, under this assumption, we can add $p(\ell)$ to our analysis without altering the basic behavioral principles discussed in Section 3.1.

The above discussion was necessary to show that (g) and (h) along with the other equilibrium conditions of GSPEM can be placed into a variational inequality form. This variational inequality form is given in the following theorem.

Theorem 3.4

The vector $\hat{x} = (\hat{S}; \hat{D}; \hat{f}; \hat{e})^T \varepsilon \Omega$ obeys the equilibrium conditions of GSPEM stated in conditions (a) - (c) and (g) - (h) if and only if:

$$\sum_{\ell \varepsilon L} \Psi_\ell(\hat{S})(S_\ell - \hat{S}_\ell) - \sum_{\ell \varepsilon L} \Theta_\ell(\hat{D})(D_\ell - \hat{D}_\ell)$$

$$+ \sum_{a \varepsilon A} \left[\sum_{v \varepsilon V} \chi_{a,v} (\hat{R}_v + \Phi\hat{t}_v - \hat{MC}^*_v) \right] (f_a - \hat{f}_a)$$

$$+ \sum_{b \varepsilon B} MC_b(\hat{e}) (e_b - \hat{e}_b) \geq 0 \qquad (3.60)$$

for all $y = (S; D; f; e) \varepsilon \Omega$.

<u>Proof</u>

Let us first prove necessity. From condition (h), for all paths $p \varepsilon P_{(i,j)}$, <u>including</u> the path $p(i) = P_{(i,i)}$, we have at equilibrium that

$$\hat{TC}_p \geq \Theta_j(\hat{D}) - \Psi_i(\hat{S}) \quad \forall\ i,j \varepsilon L,\ (i,j) \varepsilon W,\ p \varepsilon P_{(i,j)} \qquad (3.61)$$

Because this inequality is binding when $(h_p - \hat{h}_p) < 0$ ($h_p - \hat{h}_p < 0$ implies $\hat{h}_p > 0$ which implies that $\hat{TC}_p = \Theta_j(\hat{D}) - \Psi_i(\hat{S})$), we may write

$$\sum_{i \varepsilon L} \sum_{j \varepsilon L} \sum_{(i,j) \varepsilon W} \sum_{p \varepsilon P_{(i,j)}} \hat{TC}_p (h_p - \hat{h}_p) \geq \sum_{i \varepsilon L} \sum_{j \varepsilon L} \sum_{(i,j) \varepsilon W} \sum_{p \varepsilon P_{(i,j)}}$$

$$(\Theta_j(\hat{D}) - \Psi_i(\hat{S}))(h_p - \hat{h}_p) \ . \qquad (3.62)$$

Since (3.54) - (3.55) are completely equivalent to (3.53), we can rewrite the right-hand side of (3.62) by using these relationships to yield

$$\sum_{i \varepsilon L} \sum_{j \varepsilon L} \sum_{(i,j) \varepsilon W} \sum_{p \varepsilon P_{(i,j)}} (\Theta_j(D) - \Psi_i(S))(h_p - h_p)$$

$$= \sum_{\ell \varepsilon L} \Theta_\ell(\hat{D})(D_\ell - \hat{D}_\ell) - \sum_{\ell} \Psi_\ell(\hat{S})(S_\ell - \hat{S}_\ell) \ . \qquad (3.63)$$

The left-hand side of (3.62) becomes, using the relationships in Ω, (3.14), the fact that $TC_{p(\ell)} = 0 \quad \forall\ \ell = L$, and the definitions of arc costs (3.13)

$$\sum_{i\in L} \sum_{j\in L} \sum_{(i,j)\in W} \sum_{p\in P_{(i,j)}} \hat{T}C_p(h_p - \hat{h}_p)$$

$$= \sum_{i\in L} \sum_{j\in L} \sum_{(i,j)\in W} \sum_{p\in P_{(i,j)}} \hat{T}C_p(h_p-\hat{h}_p) + \sum_{i\in L} 0 \cdot (h_{p(i)}-\hat{h}_{p(i)})$$

$$= \sum_{a\in A} \hat{T}C_a(f_a - \hat{f}_a)$$

$$= \sum_{a\in A} [\sum_{v\in V} \chi_{a,v} (\hat{R}_v + \Phi\hat{t}_v)] (f_a - \hat{f}_a) . \qquad (3.64)$$

Therefore, combining (3.63) and (3.64), we can write (3.62) as

$$\sum_{\ell\in L} \Psi_\ell(\hat{S})(S_\ell - \hat{S}_\ell) - \sum_{\ell\in L} \Theta_\ell(\hat{D})(D_\ell - \hat{D}_\ell)$$

$$+ \sum_{a\in A} [\sum_{v\in V} \chi_{a,v} (\hat{R}_v + \Phi\hat{t}_v)](f_a - \hat{f}_a) \geq 0 . \qquad (3.65)$$

From conditions (a), we have at equilibrium that

$$MC_q(\hat{g}) \geq MC_v^* \quad \forall \quad v\in V, q\in Q_v \quad . \qquad (3.66)$$

Because this inequality is binding when $(g_q-\hat{g}_q) < 0$ $(g_q - \hat{g}_q < 0$ implies $\hat{g}_q > 0$ which implies that a $MC_q(\hat{g}) = MC_v^*)$, we may write

$$\sum_{b\in B} MC_b(\hat{e})(e_b - \hat{e}_b) - \sum_{v\in V} \hat{M}C_v^*(\tau_v - \hat{\tau}_v) \geq 0 ,$$

or

$$\sum_{b\in B} MC_b(\hat{e})(e_b-\hat{e}_b) - \sum_{a\in A} (\sum_{v\in V} x_{a,v} \hat{M}C_v^*)(f_a-\hat{f}_a) \geq 0. \qquad (3.67)$$

Combining (3.65) with (3.67) yields (3.60), thus proving necessity.

To prove sufficiency, assume that (3.60) obtains. By the appropriate substitutions from Ω, (3.60) can be rewritten as

$$\sum_{i\varepsilon L}\sum_{j\varepsilon L}\sum_{(i,j)\varepsilon W}\sum_{p\varepsilon P_{(i,j)}} \hat{T}C_p(h_p-\hat{h}_p) + \sum_{v\varepsilon V}\sum_{q\varepsilon Q_v} MC_q(\hat{g})(g_q-\hat{g}_q)$$

$$+ \sum_{i\varepsilon L}\sum_{j\varepsilon L}\sum_{(i,j)\varepsilon W}\sum_{p\varepsilon P_{(i,j)}} (\Psi_i(\hat{S}) - \Theta_j(\hat{D}))(h_p - \hat{h}_p)$$

$$- \sum_{v\varepsilon V}\sum_{q\varepsilon Q_v} \hat{M}C_v^{*} (g_q - \hat{g}_q) \geq 0 ,$$

or

$$\sum_{i\varepsilon L}\sum_{j\varepsilon L}\sum_{(i,j)\varepsilon W}\sum_{p\varepsilon P_{(i,j)}} \lfloor \hat{T}C_p - \Theta_j(\hat{D}) + \Psi_i(\hat{S})\rfloor h_p + \sum_{v\varepsilon V}\sum_{q\varepsilon Q_v}$$

$$\lfloor MC_q(\hat{g}) - \hat{M}C_v^{*}\rfloor g_q \geq$$

$$\sum_{i\varepsilon L}\sum_{j\varepsilon L}\sum_{(i,j)\varepsilon W}\sum_{p\varepsilon P_{(i,j)}} \lfloor \hat{T}C_p - \Theta_j(\hat{D}) + \Psi_i(\hat{S})\rfloor \hat{h}_p + \sum_{v\varepsilon V}\sum_{q\varepsilon Q_v}$$

$$\lfloor MC_q(\hat{g}) - \hat{M}C_v^{*}\rfloor g_q . \tag{3.68}$$

We see from the feasible set Ω that $S = D = h = f = g = e = 0$ is a feasible solution. Since (3.68) must hold for all feasible h and g, this implies that

$$\sum_{i\varepsilon L}\sum_{j\varepsilon L}\sum_{(i,j)\varepsilon W}\sum_{p\varepsilon P_{(i,j)}} \lfloor \hat{T}C_p - \Theta_j(\hat{D}) + \Psi_i(\hat{S})\rfloor \hat{h}_p + \sum_{v\varepsilon V}\sum_{q\varepsilon Q_v}$$

$$\lfloor MC_q(\hat{g}) - \hat{M}C_v^{*}\rfloor \hat{g}_q \leq 0 . \tag{3.69}$$

By definition, $\lfloor MC_q(\hat{g}) - \hat{M}C_v^{*}\rfloor$ is nonnegative for all $q\varepsilon Q$. To show that $\lfloor \hat{T}C_p - \Theta_j(\hat{D}) + \Psi_i(\hat{S})\rfloor$ is nonnegative for each $p\varepsilon P$, let $h = \hat{h}$ except in the one component h_p. Set $h_p = \hat{h}_p + \varepsilon$, where ε is a small positive number. This implies by the constraints in Ω that we will need to increment one path $q'\varepsilon Q_v$ for each carrier O-D pair v which the shipper path p traverses. Choose $q'\varepsilon Q_v$ for each of these carrier O-D pairs such that $\hat{M}C_{q'} = \min \{\hat{M}C_q\} = \hat{M}C_v^{*}$. Thus, set $g = \hat{g}$ except when v is traversed by shipper path p, then set $g_{q'} = \hat{g}_{q'} + \varepsilon$, where $q' \varepsilon Q_v$ is such that $[\hat{M}C_q - \hat{M}C_v^*] = 0$. Also, set $S_i = S_i + \varepsilon$ for the origin node i of shipper path $p \varepsilon P_{(i,j)}$, and set $D_j = \hat{D}_j + \varepsilon$ at the destination node j. For all other components of S and D, set $S_\ell = \hat{S}_\ell$ and $D_\ell = \hat{D}_\ell$. This new flow pattern is feasible. Therefore, (3.68) must hold for this new flow pattern. Calculating (3.68) for this flow pattern, we have

$$[\hat{T}C_p - \theta_j(\hat{D}) + \Psi_i(\hat{S})]\varepsilon \geq 0$$

or

$$\hat{T}C_p - \theta_j(\hat{D}) + \Psi_i(\hat{S}) \geq 0 \ . \qquad (3.70)$$

Since each component of the sum in (3.69) is nonnegative, the sum must equal zero, and it must be the case that each component equals zero, or that

$$[\hat{T}C_p - \theta_j(\hat{D}) + \Psi_i(\hat{S})]\hat{h}_p = 0 \quad \forall i,j \varepsilon L,\ (i,j) \varepsilon W,\ p \varepsilon P_{(i,j)} \ ,$$

$$[MC_q(\hat{g}) - \hat{M}C_v^*]\hat{g}_q = 0 \qquad \forall\ v \varepsilon V,\ q \varepsilon Q_v \ .$$

Conditions (g) and (c) are satisfied by $x \varepsilon \Omega$, and condition (b) is satisfied by use of the proper rate function. The first set of equations above imply that

$$\text{if } h_p > 0, \text{ then } \hat{T}C_p - \theta_j(\hat{D}) + \Psi_i(S) = 0$$

$$\text{if } \hat{T}C_p - \theta_j(\hat{D}) - \Psi_i(\hat{S}) > 0, \text{ then } h_p = 0,$$

or that condition (h) is satisfied. The second set of conditions imply that

$$\text{if } \hat{g}_q > 0, \text{ then } MC_q(\hat{g}) - \hat{MC}_v^* = 0$$

$$\text{if } MC_q(g) - \hat{M}C_q^* > 0, \text{ then } g_q = 0,$$

or that condition (a) is satisfied. Thus, conditions (a) - (c) and (g) - (h) are satisfied by x and sufficiency is proven.

Q.E.D.

The variational inequality just presented is not as general as the complementarity formulation in Section 3.3 due to the assumption of invertibility of the supply and demand functions. These functions are invertible if they are strictly monotone (a one-to-one mapping). The complementarity formulation does not require the functions to be monotone. However, this variational inequality formulation does yield a solution algorithm which is capable of solving large problems, as we shall see in Chapter 5. Thus, the significance of Theorem 3.4 is that we can cast GSPEM in a mathematical form which will prove useful when applying this model to realistic problems.

Finally, a special case of the variational inequality presented in Theorem 3.4 is an equivalent optimization problem (EOP) formulation of GSPEM in which many readers should be mathematically comfortable. With this EOP, the familiar Kuhn-Tucker analysis can be performed to show that a solution of this EOP emits the equilibrium conditions of GSPEM.

Let us assume that rates equal marginal costs, or $R_v = MC_v^* \ \forall \ v \varepsilon V$. Also, let us assume that the shipper time delays are exogenously aggregated, or in other words, that the time delay on shipper arc $a \varepsilon A$ can be written as a function of the flow on the arc, $\tilde{t}_a = \tilde{t}_a(f) \ \forall \ a \varepsilon A$. Finally, let us assume that all of the functions in the variational inequality are separable; that is, the functions are solely dependent upon their own variable. Under these assumptions, the variational inequality in Theorem 3.4 becomes:

$$\sum_{\ell\in L} [\Psi_\ell(\hat{S}_\ell)(S_\ell - \hat{S}_\ell) - \Theta_\ell(\hat{D}_\ell)(D_\ell - \hat{D}_\ell)]$$

$$+ \sum_{a\in A} \Phi\, \tilde{t}_a(\hat{f}_a)(f_a - \hat{f}_a)$$

$$+ \sum_{b\in B} MC_b(\hat{e}_b)(e_b - \hat{e}_b) \geq 0 \,. \qquad (3.71)$$

The variational inequality (3.71) is a statement of the first-order necessary conditions of a minimization problem (see p. 15 of Kinderlehrer and Stampacchia, 1980). This minimization problem is:

$$\text{minimize} \sum_{\ell\in L} \int_0^{S_\ell} \Psi_\ell(s)ds - \int_0^{D_\ell} \Theta_\ell(s)ds + \sum_{a\in A} \int_0^{f_a} \Phi\tilde{t}_a(s)ds$$

$$+ \sum_{b\in B} \int_0^{e_b} MC_b(s)ds \qquad (3.72)$$

subject to

$$D_\ell - S_\ell + \sum_{j\in L} \sum_{(\ell,j)\in W} \sum_{p\in P_{(\ell,j)}} h_p - \sum_{(i,\ell\in W} \sum_{p\in P_{(i,\ell)}} h_p = 0$$

$$(\pi_\ell) \quad \forall\ \ell\in L;$$

$$\sum_{a\in A} \sum_{p\in P} \chi_{a,v}\, \delta_{a,p}\, h_p - \sum_{q\in Q_v} g_q = 0 \qquad (MC^*_v) \quad \forall\ v\in V;$$

$$S_\ell,\ D_\ell \geq 0 \qquad \forall\ \ell\in L;$$

$$h_p \geq 0 \qquad \forall\ p\in P;$$

$$g_q \geq 0 \qquad \forall\ q\in Q;$$

and the definitional constraints

$$f_a = \sum_{p \varepsilon P} \delta_{a,p} h_p \qquad \forall \ a \varepsilon A;$$

$$e_b = \sum_{q \varepsilon Q} \lambda_{b,q} g_q \qquad \forall \ b \varepsilon B.$$

The above constraints are a condensed form of those constituting the feasible set Ω of Theorem 3.4, and the variables π_ℓ and MC_v^* are the dual variables of their associated constraints.

Looking first at the Kuhn-Tucker conditions of (3.72) in terms of S_ℓ and D_ℓ, we have

$$(\Psi_\ell(S_\ell) - \pi_\ell)S_\ell = 0 \tag{3.73}$$

$$\Psi_\ell(S_\ell) - \pi_\ell \geq 0, \ S_\ell \geq 0$$

and

$$(\pi_\ell - \Theta_\ell(D_\ell))D_\ell = 0 \tag{3.74}$$

$$\pi_\ell - \Theta_\ell(D_\ell) \geq 0, \ D_\ell \geq 0 .$$

Using the relation that

$$\frac{\partial}{\partial h_p} = \sum_{a \varepsilon A} \delta_{a,p} \frac{\partial}{\partial f_a} ,$$

the Kuhn-Tucker conditions in terms of h_p, $p \varepsilon P_{(i,j)}$ are

$$(\underline{MC}_p + \Phi t_p - \pi_j + \pi_i)h_p = 0 \tag{3.75}$$

$$\underline{MC}_p + \Phi t_p - \pi_j + \pi_i \geq 0, \ h_p \geq 0 ,$$

where

$$\Phi t_p = \sum_{a \varepsilon A} \delta_{a,p} \Phi t_a$$

and

$$\underline{MC}_p = \sum_{a \varepsilon A} \delta_{a,p} \sum_{v \varepsilon V} \chi_{a,v} MC_v^* .$$

Finally, noting that

$$\frac{\partial}{\partial g_q} = \sum_{b\varepsilon B} \lambda_{b,q} \frac{\partial}{\partial e_b} ,$$

the Kuhn-Tucker conditions in terms of g_q, $q\varepsilon Q_v$ are

$$(MC_q - MC_v^*)\, g_q = 0 \tag{3.76}$$

$$MC_q - MC_v^* \geq 0, \; g_q \geq 0 \; .$$

Starting with the last set of conditions (3.76), the reader can see that these conditions are a direct statement of the carrier cost minimization equilibrium condition (a). Thus, MC_v^* has the interpretation of being the minimum marginal cost of producing a move on carrier O-D pair $v\varepsilon V$. It then follows that $\underline{MC}_p$ is the marginal cost of a move on shipper path p, and by the assumption that the transportation market is purely competitive, this equals the rate charged for this service. Therefore, $\underline{MC}_p + \Phi t_p = TC_p$, the total transportation cost on shipper path p. Condition (3.75) is thus equivalent to the shipper equilibrium condition (h).

Finally, condition (3.73) implies that if there is positive supply $S_\ell > 0$, then the inverse demand function equals π_ℓ, and thus π_ℓ is equal to the market price. If $S_\ell = 0$, then the price the suppliers want, Ψ_ℓ, is greater than or equal to the prevailing market price. Likewise, if there is positive demand, $D_\ell > 0$, then the inverse demand function will equal the prevailing market price. If $D_\ell = 0$, then the market price is greater than or equal to the price which the consumers are willing to pay, Θ_ℓ.

As a final note, we observe that if all functions in (3.71) are strictly monotone then (3.72) is a strictly convex function and the Kuhn-Tucker conditions are both necessary and sufficient for a solution to (3.72). However, if these functions are non-monotonic (which implies that (3.72) may be nonconvex), we can still insure that a local minima to (3.72), which is in turn a solution to the VIP (3.71), is found.

In summary, the Kuhn-Tucker analysis has shown that a solution to (3.72) yields the equilibrium conditions for GSPEM. Harker (1984) illustrates how (3.72) can be considered a generalization of the Samuelson (1952) extremal formulation of the spatial price equilibrium concept. Basically, by adding on the last summation in (3.72) and the appropriate constraints, the extremal formulation of the spatial price equilibrium concept can be easily extended to include a model of the competitive transportation market.

Chapter 4
THE DEFINITION OF FREIGHT RATES IN GSPEM

The model which was developed in the previous chapter uses the concept of a rate reaction function to capture the carriers' rate-making behavior. This approach is necessary if we are to keep the same mathematical structure used in GSPEM because it allows us to treat each agent as operating independently of the other agent's decisions or strategies. However, if this approach is not adequate due to the need to understand the strategic responses of various agents, then much more complicated mathematical models along the lines of that by Fisk and Boyce (1983) need to be developed. This chapter focuses on the various ways in which the rate reaction function can be defined, critically evaluating them and pointing out where future modeling efforts may be needed.

Network models of the freight transportation system have been widely used in policy analysis. The behavior of the shippers, those who demand transportation services, and of the carriers, the firms which provide this service, is implicit in any discussion of the freight system. For a review of various freight network modeling techniques, the interested reader is referred to Friesz, Tobin and Harker (1983). At the core of such a model, there must be some basic sub-model of the carriers' rate-making behavior. That is, one of the basic variables to be discussed in a model of the freight transportation system is the rate charged by a carrier for a move between two points on his network. This chapter reviews the previous attempts at modeling freight rates, suggests new approaches based on recent advances in economic theory, and shows how these approaches can be used in GSPEM.

What is the problem associated with defining freight rates in GSPEM? The problem centers on knowing the demand function for the services (goods) the carriers are providing. Except for the case of pure competition, all classical models of the market behavior of a firm (monopoly, oligopoly, etc.) assume that the firm knows the demand-response to its actions. In GSPEM, the demand for service between a carrier origin-destination (O-D) pair is not given as an analytic function, but is implicitly derived by other forces which result from the assumed shippers' behavior. That is, the demand for service on a carrier's O-D pair is derived by the equilibrium behavior on the part of the shippers.

As an example of this implicitly derived demand behavior, let us assume, as in GSPEM, that the shipper chooses a sequence of carriers to move his goods. That is, to move a load from A to B, the shipper chooses carrier 1 to move the load from A to C, carrier 2 from C to D, and carrier 3 from D to B. Furthermore, the shipper is assumed to choose this path of carriers by a spatial price equilibrium process. Therefore, the demand for service on carrier 1's O-D pair A-C is not known explicitly, but is derived via the equilibrium process associated with the shippers.

Therefore, we cannot, in modeling freight rates in this network structure, rely on simple models of firm behavior due to the difficulty in defining demands. The first part of this chapter will review the previous ways in which researchers attempted to model freight rates. The next section discusses the naive approach, Section 4.2 describes the econometric approach, and Section 4.3 presents what we will call the legal-restriction approach. Next we turn to, in Sections 4.4 and 4.5, approaches to modeling freight rates which arise out of economic theory. The former presents the approaches from classical economic theory, and the latter section presents recent results in the economics literature which we will call approaches from contestable market theory. The chapter ends with a discussion on the merits and problems with each of the above approaches, and future research directions needed to perfect our definition of freight rates in GSPEM.

4.1 The Naive Approach

The simplest and most widely used approach to modeling freight rates is to assume that the rate equals some pre-specified value. That is, in solving a model of the freight transportation system, the rate is held fixed at some value. This simple representation of the rate-setting mechanism is what we will call the naive approach.

The naive approach was extremely useful when the Interstate Commerce Commission (ICC) had an almost total control over the rates charged for various types of freight movements. The rates which the ICC allowed were distance-based, and thus it was easy to find a good pre-specified value of the rate which was charged for a freight movement. However, as deregulation started in the freight transportation industry, the rates became more variable and did not follow, in general, the ICC rate formulas. Therefore, as the freight transportation industry moves in the direction of operating in a deregulated market, the naive approach will become (i) more difficult to implement, and (ii) less realistic in that the market forces which shape the structure of rates are not taken into account.

4.2 The Econometric Approach

The econometric approach to modeling freight rates consists of estimating the rate charged for a movement as a function of the distance of the move, the costs incurred in making this move, the shipment size, etc. This approach does not attempt to explicitly include the market forces at work when a carrier makes a rate decision, but rather estimates a function which uses as its data the results of such a decision.

One of the most thorough studies of freight rates using the econometric approach is the one done at MIT, the results of which are summarized in the master's thesis of Ralph Samuelson (1977). In Samuelson's work, rate functions are estimated using waybill sample data for rail, truck and barge, and for various commodity types. He finds that both miles hauled and shipment size are significant in the estimation of rates. Also, he finds empirically that 'value-of-service pricing,' whereby higher valued goods are moved with higher rates, is a supportable hypothesis in his data sample.

Many other studies of rates using this econometric approach have also been performed. The results of these studies, however, are dated in that the data used in estimating the rate function comes from the era when freight transportation was heavily regulated. If we wish to include rates in a predictive model of the freight transportation system, we need to know how rate-making occurs in a deregulated environment. Unfortunately, the data from this deregulated environment is not yet rich enough to permit good econometric studies.

Thus, the econometric approach is a good method by which rate functions can be incorporated into GSPEM, but the data necessary to implement this approach is not yet rich enough to allow for good estimations.

4.3 The Legal-Restriction Approach

Without the proper data to estimate rate functions in a deregulated environment, alternative approaches need to be developed. This section describes the legal-restriction approach in which the rates are assumed to be equal to some set of legal limits.

The Stagger's Act of 1980 (see U. S. House of Representatives, 1980) was enacted by Congress "... to allow, to the maximum extent possible, competition and the demand for services to establish reasonable rates for transportation by rail" (Section 101). However, this act did not allow unrestricted rates. In Section 201, the Act states that a rail carrier is free to charge any rate as long as this carrier does not have market dominance over the transportation market to which a particular rate applies. Market dominance is defined in Section 202 to occur when a carrier has a revenue to average variable cost percentage ratio in the market under consideration greater than the following percentages:

160% before 9-30-81,
165% between 10-1-81 and 9-30-82,
170% between 10-1-82 and 9-30-83,
175% between 10-1-83 and 9-30-84,
180% after 10-1-84.

Thus, after October 1984, a rail carrier can charge up to 1.8 times the average variable cost of producing a service and remain unregulated. Above 180%, the ICC must approve the rate.

Other legislation, both federal and state, may also contain legal restrictions on the rate-setting behavior of rail carriers and the other modes of intercity freight movement.

Therefore, the legal-restriction approach would assume that carriers price up to the legal limits, and thus the model would assume the rates are equal to these limits. The problem with this approach is its lack of realism in that carriers may have an incentive to price below the legal limits if greater profit could be derived in doing so. Also, there may not exist enough legal restrictions in some future period which we wish to forecast to make this technique operational.

4.4 Approaches from Classical Economic Theory

The approaches to modeling freight rates which have been discussed in the previous sections all rely on some type of a priori specification of the rates. That is, the rate must either be specified by some specific value or by some known functional relationship before these techniques can be used. As was pointed out, however, this type of approach to modeling freight can either be very data intensive or only possible with some assumptions which may be over-simplifications of the freight system.

In this section and the next, we describe how the results of economic theory can be applied to the problem of modeling freight rates. The approaches to freight rate modeling based on economic theory are derived from underlying assumptions about the behavior of the individual agents (carriers and shippers) comprising the freight transportation system. That is, instead of making assumptions on the rates, the approaches derived from economic theory view the rates as resulting from some underlying assumptions about the behavior of the agents.

In this section, the classical economic models of pure competition, monopolies, and oligopolistic competition will be applied to the freight rate question. Let us begin with the purely competitive model.

A purely competitive economic market is one in which each firm in the market perceives demand as being infinitely inelastic when making supply decisions (pp. 82-83 of Samuelson, 1947). In other words, each firm takes the price 'as given' when making supply decisions since it perceives its influence on the total market as being negligible (the many firms assumption). It should be noted that pure competition, unlike perfect competition, says nothing about the long-run behavior of a firm. Thus, pure competition allows negative profits to accrue to a firm in the short-run, since the only assumption made concerns the perception the firm has about demand, and nothing is said about entry and exit from the market.

Let

V^k = the set of O-D pairs under carrier k's control,

R_v = the rate charged for a move between carrier O-D pair v,

τ_v = the amount of service supplied between O-D pair v,

τ^k = $(\tau_v | v \in V^k)$, and

$\underline{C}^k(\tau^k)$ = the cost of producing the vector of outputs (services) τ^k by carrier k.

We assume that each O-D pair v is defined not only by the origin and destination nodes, but also by the level of service attributes associated with a move over this O-D pair. Thus, for each physical O-D pair, multiple outputs are defined in terms of these levels of service attributes, such as time delay, reliability, etc.

The carrier is assumed to maximize profits, this problem being denoted as:

$$\text{maximize} \sum_{v \in V^k} R_v \tau_v - C^k(\tau^k) \tag{4.1}$$

$$\text{subject to } \tau_v \geq 0 \quad \forall\, v \in V^k .$$

Taking the rate 'as given,' the carrier will maximize (4.1) by supplying τ_v up to the point where $R_v = \partial C^k / \partial \tau_v$. That is, pure competition results in marginal cost pricing.

Therefore, if the carriers serving a market are numerous enough so that each carrier views the demand as infinitely inelastic, marginal cost pricing would result. In the freight transportation market, this 'large numbers' assumption is often not reasonable. However, if it is, the resulting freight rate behavior is well-defined in GSPEM due to the fact that the demand function has never explicitly entered into formulating this rate-setting behavior.

At the other extreme from pure competition is the assumption of a monopoly, where only one carrier serves a transportation market. In this case, the carrier does not face an infinitely inelastic demand curve, but rather the total demand function for the market in question. That is, the demand for service between an O-D pair v, τ_v, is a function of the rate the carrier will charge for that service $\tau_v = \tau_v(R_v)$. Inverting this function, we can write $R_v = R_v(\tau_v)$. The carrier's profit maximization problem becomes

$$\text{maximize} \sum_{v \in V^k} R_v(\tau_v)\, \tau_v - C^k(\tau^k) \tag{4.2}$$

$$\text{subject to } \tau_v \geq 0 \quad \forall\, v \in V^k .$$

The first order conditions are, for $\tau_v > 0$,

$$R_v + \tau_v \, \partial R_v(\tau_v)/\partial\tau_v = \partial C^k/\partial\tau_v ,$$

or marginal revenue equals marginal cost.

The problem in applying (4.3) to GSPEM is that we do not know explicitly the demand function, and thus we do not know $\partial R_v(\tau_v)/\partial\tau_v$. Therefore, to assume that a carrier acts according to the classical monopoly story, we must somehow derive the value of $\partial R_v(\tau_v)/\partial\tau_v$.

The first approach which can be taken is to assume some value for $\partial R_v(\tau_v)/\partial\tau_v$. This would correspond to assuming a value for the carrier's perception of the elasticity of demand for service with respect to price (rate) between his O-D pair v. The elasticity of demand is given by

$$\varepsilon_v = \frac{\partial R_v(\tau_v)}{\partial\tau_v} \frac{\tau_v}{R_v} , \tag{4.4}$$

and thus (4.3) becomes

$$R_v(1 + \varepsilon_v) = \partial C^k/\partial\tau_v .$$

The second approach involves using the mathematical model of shippers' behavior to approximate ε_v. That is, the model of shippers' behavior can be run twice, once with R_v^1 and another time with $R_v^2 \neq R_v^1$. Then, ε_v can be approximated as

$$\varepsilon_v \approx \frac{R_v^1 - R_v^1}{\tau_v^2 - \tau_v^1} \frac{\frac{1}{2}(\tau_v^2 + \tau_v^1)}{\frac{1}{2}(R_v^2 + R_v^1)} . \tag{4.6}$$

In either of the two approaches, we cannot circumvent having to do some type of approximation to the demand function. Since the demand function is not known explicitly in GSPEM, an approximation is the best we can ever do. If this approximation is not accurate enough, then we must devise a new model along the lines of that developed by Fisk and Boyce (1983). However, it is not at all clear that the model by Fisk and Boyce is applicable to large-scale problems, whereas the application described in Chapter 6 shows that GSPEM is capable of solving large-scale problems. Thus, we are left with a choice between accuracy and applicability when describing monopolistic behavior.

Oligopolistic competition arises when the market under consideration is served by only a small number of carriers. In this case, the carriers not only face a downward-sloping demand function for their services between O-D pairs (not infinitely inelastic), but they also must be aware of the other carriers' actions. That is, to model an oligopolistic market, we need to know not only the elasticity of demand for each carrier serving that market, but we also must know the strategic response of the carriers to the other carrier's actions.

The need to know strategic responses was the impetus for the beginnings of game theory. The book by Friedman (1977) chronicles the development of game theoretic approaches to modeling oligopolistic markets. Concepts such as Nash-Cournot equilibria, Stackelberg leader-follower models, reaction functions, etc., all derive from researchers attempting to understand oligopolistic markets.

We need only note that by assuming oligopolistic competition, the amount of added approximations to elasticities, strategic behaviors, etc. make this a very difficult concept to incorporate into GSPEM. A carrier's own-rate elasticities of demand, cross-rate elasticities with other carriers, etc. must all be approximated.

The most powerful result which we can obtain concerning the question of defining freight rates in GSPEM is derived from the assumption of pure competition. Pure competition assumes nothing about the total demand function for a particular market. It only assumes that each carrier perceives the demand function he faces as being infinitely inelastic. The assumption of many firms servicing a particular market would yield, at least approximately, a highly inelastic demand function for each carrier.

As we move towards more complicated market behavior, more information in the form of a demand function is necessary. However, this demand function is difficult to derive in GSPEM. Therefore, we must resort to specifying by some means various parameters (own-rate elasticities, cross-rate elasticities, etc.) of the demand function. Thus, the implementation of the monopoly and oligopoly models from classical economic theory have embedded in them an element of the naive approach.

Let us now turn to some more recent results from contestable market theory.

4.5 Approaches from Contestable Market Theory

Economic theorists have always recognized in some way the pressure which the potential entrants into a market can place on the incumbent firms. Potential entry causes the incumbent firms not to grossly overprice their goods, since if they did, potential entrants could come into the market and underprice the incumbents. The explicit treatment of this pressure and its implications on market behavior and industrial organization is the focus of the work done by Baumol, Panzar and Willig (1982), which they call contestable market theory.

What contestable market theory brings to the problem of defining freight rates is more information. The information this theory adds is the behavior of the carriers in response to the external pressures brought on by the potential carriers depicted in Figure 2.1. Let us explore the concept of contestable markets and the relevant theorems on the market/price structure in some detail.

A perfectly contestable market is one in which potential entrants have the capability of entering and which has the following two properties: First, the potential entrants can serve the same market demand and can use the same production technologies as those which are available to the incumbent firms, without restriction. Second, the potential entrant, when contemplating entry, evaluates the profitability of entry at the prices facing the incumbent firms in the market (see p. 5 of Baumol, Panzar and Willig, 1982). Therefore, in a perfectly contestable market, entry is possible and the means of production are available to any potential entrant.

Next Baumol, Panzar and Willig (1982, p. 5) define the concept of a sustainable industry configuration:

> ... a price vector and a set of output vectors, one for each of the firms in the configuration, with the following properties: First, the quantities demanded by the market at the prices in question must equal the sum of the outputs of all the firms in the configuration [or, the total supply]. Second, the prices must yield to each active firm revenues that are no less than the cost of producing its outputs, and, last, there must be no opportunities for entry that appear profitable to potential entrants who regard the prices of the incumbent firms as fixed.

Thus, sustainable industry configurations are a type of equilibrium concept. When the market is contestable, then sustainability is necessary for an equilibrium (pp. 10-11).

The relevant theorems which Baumol, Panzar and Willig (1982) prove are stated in the following propositions, which are numbered according to the scheme in their book:

Proposition 11B1 In a sustainable industry configuration, each firm must (i) operate efficiently [or, minimize the cost of producing the vector of outputs], (ii) earn zero economic profit, (iii) avoid cross subsidies [between various output markets which it serves], (iv) select prices for each of its products which are at least as large as marginal costs, and (v) select an output vector at which [the average cost of producing this output vector (ray average costs)] is no larger than it is at any proportionately smaller output vector.

<u>Proposition 11B5</u> If two or more firms produce a given good in a sustainable configuration, then they must all select outputs at which their marginal costs of producing it are equal to one another and to the good's market price.

Therefore, in a contestable market which has a sustainable industry configuration, if two or more firms produce an output, then price equals marginal cost for both these firms. Also, for an equilibrium to exist in a perfectly contestable market, the market must have a sustainable configuration.

How then can these concepts be used to help define freight rates in GSPEM? Let us begin our discussion by briefly reviewing some basic characteristics of GSPEM.

GSPEM assumes that a short-run analysis is to be performed due to the fixed infrastructure which comprises the network. That is, the fixed capital which comprises the railways, highways, and waterways is not 'altered' during a run of GSPEM, and thus, long-run analyses cannot be done. Furthermore, each carrier in GSPEM is represented by the network on which he operates. If we were to model the entry and exit of carriers in the transportation market, we would have to alter (add or remove elements of) the network. Therefore, in using a network representation of the carrier's production function or process, it becomes very difficult to model entry and exit.

Thus, to use the concepts from contestable market theory, we must assume that the market which we have defined via our network is in a sustainable industry configuration. That is, before we start our analysis, we must assume that the set of carriers which comprises the freight transportation network can be sustainable without any alterations to this set. However, this assumption does <u>not</u> say that potential entry does not put pressure on the incumbent firms. The assumption says that the set of carriers has been defined so that no potential carrier finds it profitable to enter the market, and no incumbent carrier has an incentive to exit the market.

Given the assumption that we have a sustainable industry configuration, we know by Proposition 11B1 that the rate charged for an O-D move, R_V, is at least as great as the marginal cost of producing that move, MC_V^*, or $R_V \geq MC_V^*$. Also, by Proposition 11B5, we know that in this sustainable industry configuration, if two or more firms serve a particular O-D pair, then $R_V = MC_V^*$.

Therefore, if we assume in GSPEM not only that we have a sustainable industry configuration, but also that each O-D pair has two or more carriers serving it, then marginal cost rates ($R_V = MC_V^*$) result. What the contestable market theory has allowed us to do in this case is to expand the story of pure competition. The assumption of many carriers serving a particular O-D pair is no longer necessary for marginal cost pricing to result. If only a few carriers serve an O-D pair, but there is almost free entry into this market, then marginal cost pricing will also occur. For what follows, we will call a market contestable if it is perfectly contestable and has reached a sustainable configuration.

How reasonable is the assumption of free entry which underlies the concept of perfect contestability? For motor carriers, this assumption may not be a bad approximation in that trucks are relatively inexpensive and the majority of the infrastructure necessary to produce transportation service is not provided by the carrier. For rail, however, this assumption seems unrealistic due to the large capital investment which is necessary to create the means of production (the railway). Keeler (1983), in Chapter 3, also discusses the use of the results from contestable market theory in analyzing the railroad industry, and has a similar conclusion. Water carriers are very similar to motor carriers in that the majority of the infrastructure over which they operate is provided by the government. Therefore, free entry may not be a bad approximation for the water transportation market.

One barometer which can be used to test the assumption of contestability is the cross subsidy issue (item (iii) of Proposition 11B1). If a transportation market is contestable, then no carrier should be underpricing one O-D move and then overpricing another move to compensate for the underpriced one. If a large amount of cross subsidy is seen to exist, then contestability is not a good assumption to make. However, to perform this test on transportation markets, we must wait until the 'dust' settles from the recent regulatory reforms. That is, the current transportation market is in a state of change from a regulated to deregulated environment. Contestability arguments assume free market behavior. Thus, to test contestability, we must wait until the markets begin operating in something close to a fully deregulated environment.

4.6 Conclusions

This chapter has presented the various methods which can be used to define freight rate-making behavior in GSPEM. As one can see, there is no one best approach to use. The econometric approach is very robust in that many parameters such as shipment size, reliability, etc. can be easily added. However, this approach is very data intensive, and the data from the deregulated environment is not yet rich enough to permit good estimates.

The approaches arising out of economic theory also have difficulties associated with them. Models of monopolies and oligopolistic competition need a demand function to be operational, but this demand function is difficult, if not impossible, to define in the type of network model which we have been discussing. If a different type of monopolistic model such as the one by Fisk and Boyce (1983) is used, it is not clear that it can be solved for problems of realistic size. Purely competitive and contestable market approaches do not need an explicit demand function to be operational, but their underlying assumptions may not be realistic for particular applications.

Future research must be focused on this question of how carriers compete and collude with one another to set freight rates. The days of ICC regulation and its associated rate formulas are over. The behavior of a small number of agents who must often work together to produce transportation services must be better understood.

Therefore, to define the rate reaction function in GSPEM, we must currently rely on the approaches presented in this chapter. The most precise definition of the rate reaction function results from purely competitive and contestable market assumptions. From these assumptions, marginal cost pricing results without any need to define the demand function for an O-D market. As the next chapter discusses, marginal cost pricing will also yield a solution algorithm which is capable of solving very large problems. Let us now begin our discussion procedures for GSPEM.

Chapter 5
COMPUTATIONAL ALGORITHMS FOR GSPEM

In Chapter 3, the theoretical framework for the Generalized Spatial Price Equilibrium Model (GSPEM) was developed. This chapter addresses the following question: is the GSPEM operational? That is, can a solution to GSPEM be computed?

Two solution algorithms are presented in this chapter. The first algorithm presented is based upon the nonlinear complementarity formulation of GSPEM. This algorithm is capable of solving GSPEM for a general specification of the rate reaction function. The next section presents this algorithm. In Section 5.1, a solution algorithm based on the variational inequality formulation of GSPEM is presented. This algorithm assumes that rates equal marginal costs and that the time delay functions are not derived via the time delays on carrier paths, but are specified as functions of the flows on the shipper network (exogenous aggregation). Numerical examples of the two algorithms are presented in Section 5.3. Finally, several recent advances in variational inequality algorithms will be reviewed in Section 5.4 and their suitability for use in solving GSPEM will be discussed.

5.1 Iterative Complementarity Algorithm

A nonlinear complementarity problem (NCP) is to find a vector $x = (x_1, x_2, \ldots, x_n)^T$ such that for the vector function $F(x) = (F_1(x), F_2(x), \ldots, F_n(x))^T$, the following conditions hold:

$$F(x)^T \cdot x = 0 , \quad (5.1)$$
$$F(x) \geq 0,$$
$$x \geq 0 .$$

The notation which we will use throughout this chapter is the same as that used in Chapter 3, and is summarized in Appendix A. Let us begin our discussion by restating the NCP formulation of GSPEM, which is:

$$F = \lfloor S_\ell(\pi) - D_\ell(\pi) + \sum_{i \varepsilon L} \sum_{(i,\ell) \varepsilon W} \sum_{p \varepsilon P_{(i,\ell)}} h_p - \sum_{j \varepsilon L} \sum_{(\ell,j) \varepsilon W} \sum_{p \varepsilon P_{(\ell,j)}} h_p$$

$$\forall\ \ell\ \varepsilon\ L\ ;$$

$$\pi_i + TC_p - \pi_j \qquad \forall\ i,j\ \varepsilon L,\ (i,j) \varepsilon W,\ p \varepsilon P_{(i,j)}\ ;$$

$$\sum_{q \varepsilon Q_v} g_q - \sum_{a \varepsilon A} \chi_{a,v} \left(\sum_{p \varepsilon P} \delta_{a,p} h_p \right) \qquad \forall\ v\ \varepsilon\ V\ ;$$

$$MC_q - MC_v^* \qquad \forall\ v\ \varepsilon\ V,\ q\ \varepsilon\ Q_v \rfloor, \tag{5.2}$$

$$x = \lfloor \pi\ ;\ h\ ;\ MC^*\ ;\ g \rfloor . \tag{5.3}$$

The <u>iterative complementarity algorithm</u> which we will apply to the NCP defined by (5.2) and (5.3) forms at each iteration a linear complementarity problem (LCP), and solves this LCP by Lemke's algorithm (see Lemke and Howson, 1964). The statement of this algorithm for the NCP (5.1) is:

<u>Step 0</u> Choose an initial vector $x^o \geq 0$, set $k = 0$.

<u>Step 1</u> Evaluate the linearization of F at x^k to form $\tilde{F}$:
$\tilde{F}(x) = F(x^k) + \nabla F(x^k)^T \cdot (x - x^k)$,
where $\nabla F(x^k)$ denotes the Jacobian of F evaluated at x^k.

<u>Step 2</u> Solve the LCP
$\tilde{F}(x^{k+1})^T \cdot (x^{k+1}) \geq 0$.
$\tilde{F}(x^{k+1}) \geq 0$, $x^{k+1} \geq 0$.

Step 3 If $|x_i^{k+1} - x_i^k| \le \varepsilon$, a preset tolerance, $i=1, 2, \ldots, n$, stop; x^{k+1} is a solution to (5.1). Else, set $k = k+1$ and return to Step 1.

Karamardian (1971) pointed out that every complementarity problem is a variational inequality problem (VIP), which is to find $x \in \Omega \subseteq R^n$, where Ω is a feasible set, such that:

$$F(x)^T \cdot (y-x) \ge 0 \qquad \forall \ y \in \Omega . \tag{5.4}$$

Therefore, any results which apply to a VIP also apply to complementarity problems.

The algorithm just presented can thus be considered an algorithm for a VIP. In a recent paper, Pang and Chan (1982) state some conditions on F which will insure the local convergence of the above algorithm, which they refer to as Newton's method. The following lemma corresponds to Pang and Chan's Corollary 2.6:

Lemma 5.1 Let Ω be a nonempty closed and convex subset of R^n. Let $F: R^n \to R^n$ and $\hat{x}$ be the solution of (5.1). Suppose that F is continuously differentiable with $\nabla F(\hat{x})$ being positive definite. Then there exists a neighborhood of $\hat{x}$ such that if the initial iterate x^o is chosen there, the sequence $\{x^k\}$ generated by the iterative complementarity algorithm (or Newton's method) is well defined and converges to $\hat{x}$. Moreover, if ∇F is Lipschitz continuous at $\hat{x}$, i.e., if there exists a neighborhood N of $\hat{x}$ and a positive scalar γ such that for all x and y in N, $\|\nabla F(x) - \nabla F(y)\| \le \gamma\|x-y\|$ ($\|\cdot\|$ denotes the standard Euclidean norm), then $\{x^k\}$ converges quadratically to $\hat{x}$, i.e., there is a constant c such that

$$\|x^{k+1} - x\| \le c\ \|x^k - x\|^2 \quad \text{for all } k.$$

We want to apply this iterative complementarity algorithm to (5.2) - (5.3), but we do not want to enumerate the full path sets P and Q due to the potential enormity of these sets. Therefore, we want to implement this algorithm with a partial path enumeration scheme. The algorithm, when applied to (5.2) - (5.3) with this path enumeration scheme, will be called ALG1 and can be stated as:

Step 0 Choose an initial vector $(x^o) = (\pi^o; h^o; MC^{*o}; g^o)$. Also, choose an initial path set P_w^o $\forall$ $w \varepsilon W$ and an initial path set Q_v^o $\forall$ $v \varepsilon V$.

Step 1 For each $w \varepsilon W$, solve for the shortest path $\bar{p}_w$ and set $P_w^{k+1} = P_w^k \cup \bar{p}_w$. For each $v \varepsilon V$, solve for the shortest path $\bar{q}_v$ and set $Q_v^{k+1} = Q_v^k \cup \bar{q}_v$.

Step 2 Linearize the function F defined by (5.2) to form the linear function $\tilde{F}(x) = F(x^k) + \nabla F(x^k)^T \cdot (x-x^k)$.

Step 3 Solve the resulting LCP by Lemke's algorithm (see Lemke and Howson, 1964).

Step 4 If

$$|\pi_\ell^{k+1} - \pi_\ell^k| \le \varepsilon_1 \qquad \forall \ \ell \ \varepsilon \ L,$$

$$|h_p^{k+1} - h_p^k| \le \varepsilon_2 \qquad \forall \ p \ \varepsilon \ P,$$

$$|MC_v^{*k+1} - MC_v^{*k}| \le \varepsilon_3 \qquad \forall \ v \ \varepsilon \ V,$$

$$|g_q^{k+1} - g_q^k| \ l \le \varepsilon_4 \qquad \forall \ q \ \varepsilon \ Q,$$

and no new paths have been added to the path sets:

$$P_w^{k+1} = P_w^k \qquad \forall \ w \ \varepsilon \ W, \text{ and}$$

$$Q_v^{k+1} = Q_v^k \qquad \forall \ v \ \varepsilon \ V,$$

then stop; x^{k+1} is a solution to (5.2) - (5.3). Else,

set k = k+1 and go to step 1.

The above algorithm generates all the necessary paths, i.e., paths with positive flow at the solution to (5.2) - (5.3). If a path should be an element of a path set by is currently omitted, it must be the case that it is a shortest path and thus will be added in the next iteration. Both Aashtiani (1979) and Bertsekas and Gafni (1980) discuss this partial path enumeration scheme.

To apply Pang and Chan's local convergence result to ALG1, let us first prove the following:

Lemma 5.2 The following conditions are sufficient for the function F in (5.2) to be such that ∇F is positive definite in the vicinity of $\hat{x}$:

(i) F is continuous and continuously differentiable,

(ii) $(S_\ell(\pi) - D_\ell(\pi))$ is strictly monotone (increasing) $\forall\ \ell \in L$,
$MC_b(e)$ is strictly monotone (increasing) $\forall\ b \in B$,
$TC_a(f)$ is strictly monotone (increasing) $\forall\ a \in A$,
and
$(R_v + \Phi t_v - MC_v^*)$ is strictly monotone (increasing) $\forall\ v \in V$,

for all x in the neighborhood of $\hat{x}$.

Proof From Ortega and Rheinboldt (1970), p. 143, we know that ∇F is positive definite if F is a strictly monotone function. Therefore, we only need to show that F is strictly monotone in the neighborhood of $\hat{x}$ to prove that ∇F is positive definite in that region.

From the proof of Theorem 3.3 we know that the strict monotonicity definition becomes for the function F defined in (5.2):

$$[F(\hat{x}) - F(x)]^T \cdot (\hat{x}-x) = \sum_{\ell \in L} [(S_\ell(\hat{\pi}) - D_\ell(\hat{\pi})) - (S_\ell(\pi) - D_\ell(\pi))]$$

$$(\hat{\pi}_\ell - \pi_\ell) + \sum_{a \in A} [TC_a(\hat{f}) - TC_a(f)] (\hat{f}_a - f_a)$$

$$+ \sum_{v \in V} [(\hat{R}_v + \Phi\hat{t}_v - \hat{MC}^*_v) - (R_v + \Phi t_v - MC^*_v)] (\hat{\tau}_v - \tau_v)$$

$$+ \sum_{b \in B} [MC_b(\hat{e}) - MC_b(e)] (\hat{e}_b - e_b) > 0 \tag{5.5}$$

for $x \neq \hat{x}$, where f, τ and e are derived from h and g. By assumption, each component function in (5.5) is strictly monotone, which implies that F is strictly monotone. Thus, ∇F must be positive definite in the vicinity of $\hat{x}$.

Q.E.D.

Given the above result, we now state the local convergence result for ALG1:

<u>Theorem 5.1</u> If the function F defined in (5.2) satisfies the conditions of Lemma 5.2, then there exists a neighborhood of $\hat{x}$, a solution to the NCP formulation of GSPEM, such that if the initial iterate x^o is chosen there, the sequence $\{x^k\}$ generated by ALG1 is well-defined and converges to $\hat{x}$.

Furthermore, if ∇F is Lipschitz continuous at $\hat{x}$, then $\{x^k\}$ converges <u>quadratically</u> to $\hat{x}$.

<u>Proof</u> If F satisfies the conditions of Lemma 5.2, then ∇F is positive definite and the result follows directly from Lemma 5.1, where $\Omega = R^n_+$.

Q.E.D.

By Theorem 3.3 of Chapter 3, we know that if F satisfies the conditions listed in Lemma 5.2, for all x in Ω then GSPEM has a unique solution. Thus, under these same conditions, ALG1 will converge quadratically to this unique solution. However, if the conditions of Lemma 5.2 are not met, we still may be able to show local convergence if $\nabla F(\hat{x})$ remains positive definite. Therefore, without uniqueness being assured, we still can show that ALG1 will converge to a solution of GSPEM if ∇F is positive definite at that solution and x^0 is chosen close enough to the solution.

As was pointed out in the uniqueness discussion in Chapter 3, strict monotonicity of the functions listed in Lemma 5.2 may be unrealistic in many freight applications due to the presence of U-shaped average cost and time delay functions. It is possible with these nonmonotonic functions to assure the convergence of ALG1 if the conditions mentioned in the preceding paragraph are met. However, in general, we may not be able to show ∇F is positive definite. Still we can apply ALG1 to the model, and be assured by the following theorem that if it converges, it converges to a true solution of the model:

Theorem 5.2 If ALG1 converges to a solution of the NCP generated by (5.2) - (5.3) in the sense that $x^{k+1} \simeq x^k$, then x^{k+1} is a solution to GSPEM.

Proof If $x^{k+1} \simeq x^k$, then
$$\tilde{F}(x^{k+1}) = F(x^k) + \nabla F(x^k)\cdot(x^{k+1} - x^k)$$
$$\simeq F(x^{k+1}) + 0 = F(x^{k+1})$$

by the continuity of F and ∇F.

Thus, at the final iteration we have

$$\tilde{F}(x^{k+1})^T \cdot (x^{k+1}) = 0,\ \tilde{F}(x^{k+1}) \geq 0,\ x^{k+1} \geq 0\ ,$$

which is approximately equivalent to

$$F(x^{k+1})^T \cdot (x^{k+1}) = 0,\ F(x^{k+1}) \geq 0,\ x^{k+1} \geq 0\ ,$$

which is the definition of a solution to the NCP generated by (5.2) - (5.3). The NCP (5.2) - (5.3) has been shown in Theorem 3.1 to be completely equivalent to GSPEM, and thus x^{k+1} must be a solution to GSPEM.

Q.E.D.

Therefore, even if we cannot assure convergence of ALG1, we know that if it does converge, it will not converge to some spurious solution, but to a true solution of GSPEM.

Pang and Chan (1982) also show in their Corollary 2.11 that if the functions listed in Lemma 5.2 are strictly monotone, then convergence can be assured for any $x^o \varepsilon \Omega$. This global convergence condition is of course even more difficult to meet in some realistic freight problems. However, if it is met, then we know that ALG1 will converge to the unique solution no matter where we start the algorithm.

Let us now turn to an optimization-based solution algorithm arising out of the VIP form of GSPEM when marginal cost pricing and explicit aggregation of the time delay function are assumed.

5.2 Diagonalization/Relaxation Algorithm

As was stated in the previous section, a variational inequality problem (VIP) is to

$$\text{find } x \ \varepsilon \ \Omega \qquad (5.6)$$

such that

$$F(x)^T \cdot (y-x) \geq 0 \quad \forall \quad y \ \varepsilon \ \Omega \ .$$

The VIP formulation of GSPEM derived in Theorem 3.4 is to find

$$\hat{x} = (\hat{S};\ \hat{D};\ \hat{f};\ \hat{e})\varepsilon\Omega, \text{ where}$$

$$\Omega = \{x \mid D_\ell - S_\ell + \sum_{j\varepsilon L} \sum_{(\ell,j)\varepsilon W} T_{(\ell,j)} - \sum_{i\varepsilon L} \sum_{(i,\ell)\varepsilon W} T_{(i,\ell)} = 0 \qquad \forall \ \ell \ \varepsilon \ L \ ,$$

$$T_w = \sum_{p\varepsilon P_w} h_p \qquad \forall \ w \ \varepsilon \ W \ ;$$

$$f_a = \sum_{p \in P} \delta_{a,p} h_p \qquad \forall \ a \in A ;$$

$$\tau_v = \sum_{a \in A} \chi_{a,v} f_a \qquad \forall \ v \in V ;$$

$$\tau_v = \sum_{q \in Q_v} g_q \qquad \forall \ v \in V ;$$

$$e_b = \sum_{q \in Q} \lambda_{b,q} g_q \qquad \forall \ b \in B ;$$

$$x \geq 0\}$$

such that

$$F(\hat{x})^T \cdot (x-\hat{x}) = \sum_{\ell \in L} [\Psi_\ell(\hat{S})(S_\ell-\hat{S}_\ell) - \Theta_\ell(\hat{D})(D_\ell-\hat{D}_\ell)]$$

$$+ \sum_{a \in A} \sum_{v \in V} \chi_{a,v}(\hat{R}_v + \Phi \hat{t}_v - \hat{MC}_v^*)(f_a-\hat{f}_a)$$

$$+ \sum_{b \in B} MC_b(\hat{e})(e_b-\hat{e}_b) \geq 0 \qquad (5.7)$$

for all $x = (S; D; f; e) \in \Omega$.

Let us make the following two assumptions:

(1) $R_V = MC_V^* \ \forall \ v \in V$; that is, marginal cost pricing holds for all O-D pairs. This condition can result from either an assumption of purely competitive or contestable markets (see Chapter 4).

(2) There is exogenous aggregation of the time delays; that is, the time delay on a shipper arc a are stated explicitly as a function of the flows on the shipper network, $\tilde{t}_a = \tilde{t}_a(f)$ (see Chapter 3).

Given these two assumptions, the second function in (5.7) becomes:

$$\sum_{v\varepsilon V} \chi_{a,v}(R_v+\Phi\, t_v-MC_v^*) = (\sum_{v\varepsilon V} \chi_{a,v}\, MC_v^*+\Phi\, \tilde{t}_a(f)) - \sum_{v\varepsilon V} \chi_{a,v}\, MC_v^*$$

$$= \Phi\, \tilde{t}_a(f)\ , \qquad (5.8)$$

The VIP (5.7) thus becomes

find $\hat{x} = (\hat{S};\ \hat{D};\ \hat{f};\ \hat{e})\varepsilon\ \Omega$

such that

$$F(\hat{x})^T\cdot(x-\hat{x}) \;=\; \sum_{\ell\varepsilon L} [\Psi_\ell(\hat{S})(S_\ell-\hat{S}_\ell) - \Theta_\ell(\hat{D})(D_\ell-\hat{D}_\ell)]$$

$$+ \sum_{a\varepsilon A} \Phi\, \tilde{t}_a(f)(\hat{f}_a-f_a)$$

$$+ \sum_{b\varepsilon B} MC_b(\hat{e})(e_b-\hat{e}_b) \geq 0 \qquad (5.9)$$

for all $x\ \varepsilon\ \Omega$.

This section focuses on the solution of (5.9). In the next subsection, the diagonalization algorithm for VIP's is discussed, and then applied to (5.9). Next, the subproblem which results from the application of the diagonalization algorithm is studied, and a feasible direction algorithm is developed for its solution. Finally, the entire algorithm for (5.9) is stated and conclusions are drawn.

Definition and Application of a Diagonalization Algorithm.

Many algorithms have been developed for the solution of variational inequality problems. The books by Glowinski, Lions and Tremelieres (1976) and Cottle, Giannessi, and Lions (1980) describe many of these techniques. In this section, we are concerned with what is known as a <u>diagonalization</u> (<u>or relaxation</u>

or Jacobi approximation) algorithm for the solution of a VIP. Let us briefly describe this algorithm.

The diagonalization algorithm has at its core the creation of the following function $\tilde{F}(x^{k+1}, x^k) = (\tilde{F}_1(x_1^{k+1}, x^k), \tilde{F}_2(x_2^{k+1}, x^k), \ldots, \ddot{F}_n(x_n^{k+1}, x^k))$ at each iteration k+1, where

$$\ddot{F}_i(x^{k+1}, x^k) = F_i(x_1^k, \ldots, x_{i-1}^k, x_i^{k+1}, x_{i+1}^k, \ldots, x_n^k). \quad (5.10)$$

As one can see, the Jacobian of $\ddot{F}$ is diagonal, and hence the name of this algorithm. The steps of the diagonalization are as follows:

Step 0 Choose an initial iterate $x^o \in \Omega$, set $k = 0$.

Step 1 Solve the following VIP for x^{k+1}:
find $x^{k+1} \in \Omega$
such that
$\ddot{F}(x^{k+1}, x^k)^T \cdot (x - x^{k+1}) \geq 0 \quad \forall\, x \in \Omega$.

Step 2 If $|x_i^{k+1} - x_i^k| \leq \varepsilon$ i=1, 2, ..., n, stop; x^{k+1} is a solution to the original VIP. Else, set $k = k+1$ and return to step 1.

Pang and Chan (1982), in their Theorem 5.2, have proven that the above algorithm converges locally, this result being summarized in the following lemma:

Lemma 5.3 Let $\ddot{x}$ solve the VIP (5.6). Suppose that the conditions below are satisfied:

(i) F is differentiable and
$\partial F_i(x)/\partial x_i \geq 0 \quad \forall i = 1,2,\ldots,n, \; x \in \Omega$,

(ii) Ω is a convex set,

(iii) F is continuously differentiable in the neighborhood of $\hat{x}$,

(iv) $\partial F_i(\hat{x})/\partial x_i > 0 \quad i=1,2,\ldots,n$, and

(v) $\|D^{-1/2} BD^{-1/2}\| < 1$, where D and B are respectively the diagonal and off-diagonal parts of $\nabla F(\hat{x})$ and $\|\cdot\|$ denotes the standard Euclidean norm.

Then, provided that the initial iterate x^o is chosen in the suitable neighborhood of $\hat{x}$, the sequence $\{x^k\}$ generated by the diagonalization (or Jacobi) method will converge to $\hat{x}$.

In the remark to their Corollary 2.4, p. 292, Pang and Chan state that if condition (v) holds at $\hat{x}$, then $\nabla F(x)$ must be positive definite. This in turn implies that F is a strictly monotone function <u>at $\hat{x}$</u> if (v) holds. Therefore, we have the result that:

<u>Lemma 5.4</u> A necessary condition for condition (v) of Lemma 5.3 to hold is that F be a strictly monotone function <u>at $\hat{x}$</u>.

The application of this diagonalization algorithm to (5.9) will be called ALG2, and can be stated as follows:

<u>Step 0</u> Choose an initial iterate $x^o = (S^o; D^o; f^o; e^o) \varepsilon\ \Omega$, where Ω is as defined in (5.7). Set $k = 0$.

<u>Step 1</u> Solve the following VIP for x^{k+1}:
find $x_{k+1}\ \varepsilon\ \Omega$
such that

$$\tilde{F}(x^{k+1}, x^k)^T \cdot (x - x^{k+1}) = \sum_{\ell \varepsilon L} [\Psi_\ell(S_\ell^{k+1}, S^k)(S_\ell - S_\ell^{k+1})$$

$$- \Theta_\ell(D_\ell^{k+1}, D^k)(D_\ell - D_\ell^{k+1})]$$

$$+ \sum_{a \in A} \Phi\, t_a(f_a^{k+1}, f^k)(f_a - f_a^{k+1})$$

$$+ \sum_{b \in B} MC_b(e_b^{k+1}, e^k)(e_b - e_b^{k+1}) \geq 0 \qquad (5.11)$$

for all $x \in \Omega$.

Step 2 If $|S_\ell^{k+1} - S_\ell^k| \leq \varepsilon_5 \qquad \forall\ \ell \in L$,

$|D_\ell^{k+1} - D_\ell^k| \leq \varepsilon_6 \qquad \forall\ \ell \in L$,

$|f_a^{k+1} - f_a^k| \leq \varepsilon_5 \qquad \forall\ a \in A$, and

$|e_b^{k+1} - e_b^k| \leq \varepsilon_8 \qquad \forall\ b \in B$,

then stop; x^{k+1} is a solution to (5.9). Else, set $k = k+1$ and return to step 1.

We now state the local convergence proof for ALG2:

Theorem 5.3 If, for the solution vector $\hat{x} = (\hat{S}, \hat{D}, \hat{f}, \hat{e})$,

(i) $\Psi_\ell(S)$ is continuous and continuously differentiable, with $\partial\Psi_\ell(S)/\partial S_\ell \geq 0$ for all feasible S, and Ψ_ℓ is strictly monotone at $\hat{x}$ $\forall\ \ell \in L$,

(ii) $\Theta_\ell(D)$ is continuous and continuously differentiable, with $\partial\Theta_\ell(D)/\partial D_\ell \leq 0$ for all feasible D, and $-\Theta_\ell$ is strictly monotone at $\hat{x}$ $\forall\ a \in A$,

(iii) $\tilde{t}_a(f)$ is continuous and continuously differentiable, with $\partial\tilde{t}_a(f)/\partial f_a \geq 0$ for all feasible f, and $\tilde{t}_a$ is strictly monotone at $\hat{x}$ $\forall$ $a \in A$,

(iv) $MC_b(e)$ is continuous and continuously differentiable, with $\partial MC_b(e)/\partial e_b \geq$ for all feasible e, and MC_b is strictly monotone at $\hat{x}$ $\forall$ $b \in B$,

then, provided that the initial iterate x^o is chosen in the suitable neighborhood of x and the function F defined in (5.9) is such that $\|D^{-1/2}\ B\ D^{-1/2}\| < 1$, where D and B are respectively the diagonal and off-diagonal parts of $\nabla F(\hat{x})$, the sequence $\{x^k\}$ generated by ALG2 will converge to the solution x of (5.9).

Proof First, the feasible set Ω defined by (5.7) is composed of linear functions, and thus is convex, satisfying condition (ii) of Lemma 5.3.

Conditions (i) and (iii) of Lemma 5.3 are met by the assumptions of continuity and continuous differentiability of Ψ_ℓ, Θ_ℓ, $\tilde{t}_a$, and MC_b.

Conditions (iv) and (v) of Lemma 5.3 are met by the assumptions of strict monotonicity at $\hat{x}$. Strict monotonicity implies a strictly increasing function, and thus $\partial F_i(x)/\partial x_i > 0$. Strict monotonicity and the assumption that $\|D^{-1/2}\ B\ D^{-1/2}\| < 1$ implies that condition (v) is met.

Therefore, all the conditions of Lemma 5.3 are met, and the result follows directly.

Q.E.D.

Dafermos (1983) has shown that if F is strictly monotone for all $x \in \Omega$, then global convergence can be assured. That is, strict monotonicity of F defined in (5.9) implies there is a unique solution by Theorem 3.3 in Chapter 3, and Dafermos has shown that the diagonalization algorithm will find this solution for any initial iterate $x^o \in \Omega$, provided the conditions of Lemma 5.3 are met.

As was discussed in Section 5.2, strict monotonicity of the functions listed in Theorem 5.3 may not be reasonable for some realistic freight applications. Theorem 5.3 only requires strict monotonicity at a solution, and thus local convergence might possibly be shown with U-shaped functions. However, if this cannot be done, we still know that if ALG2 converges, it converges not to some spurious solution, but to a true solution of GSPEM. This result is stated as:

Theorem 5.4 If ALG2 converges to a solution of the VIP (5.9) in the sense that $x^{k+1} \cong x^k$, then x^{k+1} is a solution to GSPEM.

Proof If $x^{k+1} \cong x^k$, then $F(x^{k+1}, x^k) \cong \tilde{F}(x^{k+1}, x^{k+1}) = F(x^{k+1})$ by the continuity of F. Thus, in step 1, we must have $F(x^{k+1})^T \cdot (x - x^{k+1}) \geq 0 \; x \in \Omega$. Therefore, x^{k+1} is a solution to (5.9), which is completely equivalent to being a solution to GSPEM.

Q.E.D.

The question which we now must address is how to solve the VIP generated in the second step of ALG2; to this question we now turn.

A Feasible Direction Algorithm for the Variational Inequality Subproblem.

There is a well-known relationship between variational inequality problems and optimization problems which is stated in the following lemma (see Proposition 5.1 on p. 15 of Kinderlehrer and Stampacchia, 1980):

Lemma 5.5 Suppose there exists an $x \in \Omega$ such that

$$Z(x) = \min_{y\in\Omega} Z(y) . \tag{5.12}$$

Then x is a solution of the variational inequality

$$x \in \Omega: F(x)^T \cdot (y-x) \geq 0 \quad \forall\, y \in \Omega ,$$

where $F(x) = \nabla Z(x)$ (= grad $Z(x)$).

Therefore, a VIP is a statement of the first-order necessary conditions of a differentiable optimization problem.

Note that in Lemma 5.5, no mention was made of convexity of $Z(x)$, or equivalently, mononoticity of $F(x)$. That is, the relationship between VIPs and the optimization problems defined in Lemma 5.5 does not depend upon monotonicity of $F(x)$. Any local minima of $Z(x)$ must satisfy the first-order necessary conditions for optimality, and thus must satisfy the VIP.

If $F(x)$ is monotone, then $Z(x)$ will be a convex function and (5.12) becomes a convex programming problem.

In this section, we will explore how the variational inequality problem in step 1 of ALG2 can be converted into an optimization problem, and then solved via a feasible direction algorithm. Let us begin by placing this VIP into an extremal formulation.

The function $\tilde{F}$ defined in step 1 of ALG2 is such that its Jacobian is diagonal, and hence symmetric. Let us integrate F to form the following problem:

$$\min_{x^{k+1}\in\Omega} \quad Z(x^{k+1}) = \oint_0^{x^{k+1}} F(x, x^k)dx , \tag{5.13}$$

where '$\oint$' denotes a line integral. We know by the symmetry of the Jacobian of F and Green's theorem that this line integral is path independent, so that (5.13) may be rewritten as:

$$\min_{x^{k+1} \in \Omega} Z(x^{k+1}) = \int_0^{x^{k+1}} F(x, x^k)dx \, ,$$

Writing $Z(x^{k+1})$ out in full, we have

$$Z(x^{k+1}) = \sum_{\ell \in L} \left[\int_0^{S_\ell^{k+1}} \Psi_\ell(s, S^k)ds - \int_0^{D_\ell^{k+1}} \Theta_\ell(s, D^k)ds \right]$$

$$+ \sum_{a \in A} \Phi \int_0^{f_a^{k+1}} t_a(f, f^k)df + \sum_{b \in B} \int_0^{e_b^{k+1}} MC_b(e, e^k)de \, .$$

$$(5.15)$$

The reason for the assumption of marginal cost pricing lies in the formation of (5.15). If the term $(R_v - MC_v^*)$ remained in the VIP (5.11), it is not clear how this term can be integrated to form (5.15). However, by assuming $R_v = MC_v^*$, integrability of the functions in (5.11) is a fairly easy condition to meet. This result is not surprising in that most analyses of economic systems assume marginal cost pricing due to the ease of computing a solution with such an assumption.

In general, Z may be a nonconvex function due to the presence of a nonmonotonic function arising in some freight applications. However, we know from the above that if x^{k+1} is a local minima of Z, then it must be a solution to the VIP defined in step 1 of ALG2. Therefore, if we can find a local minimum to the problem defined in (5.14), we have solved the subproblem defined in step 1 of ALG2. Of course, if all the functions in our problem are strictly monotone, then Z is strictly convex and there exists only one local minima (the global minimum).

To solve (5.14), we are going to apply a feasible direction algorithm. In particular, due to the presence of path variables in the feasible set Ω, we will apply the Frank-Wolfe decomposition algorithm which is widely used in urban traffic assignment problems (see LeBlanc, Morlok and Pierskalla (1975) and Gartner (1977)). This approach allows for dynamic or partial enumeration of the path sets P and Q. However, these path sets need not be stored from iteration to iteration, as in the path enumeration scheme for ALG1, and thus very substantial benefits are gained in the size of the problem which can be addressed. Let us begin with a brief review of feasible direction algorithms for general, i.e., possibly nonconvex, objective functions.

The general problem with which we are concerned is of the form

$$\min Z(x) \quad \text{subject to } x \in \Omega , \tag{5.16}$$

where Ω is of the form $Ax = b$, i.e., a polyhedral set. Following Section 13.4 of Avriel (1976), a feasible direction algorithm selects at each iteration i a feasible descent direction y^i and a step length along this feasible direction α^i such that the sequence $\{x^i\}$ generated converges to a (local) minima of Z.

There are many ways in which a feasible descent direction can be generated, but all these methods must operate such that the direction vector is bounded, i.e.,

$$\| y^i \| \leq 1 ,$$

if y^i is a normalized vector, or, in general

$$\| y^i \| < \infty . \tag{5.17}$$

The technique which we will employ here is the technique by Frank and Wolfe (1956) in which a linear approximation of the objective function Z is minimized subject to constraints that assure feasibility of the search direction. If Ω is bounded or if for each $x \in \Omega$, the set $\{y \mid \nabla Z(x)^T \cdot y < 0,\ y \in \Omega\}$ is bounded from below, then (5.17) can be ignored. However, if either of these conditions is not met, (5.17) must be included in the linear-program (LP) generated by the Frank-Wolfe technique so that the direction vector remains bounded. At each iteration i+1, the direction vector generated by the Frank-Wolfe technique is $y^{i+1} = \tilde{x}^{i+1} - x^i$, where $\tilde{x}^{i+1}$ is the solution of the LP. This LP takes the form

$$\text{minimize } \tilde{Z}(\tilde{x}^{i+1}) = Z(x^i) + \nabla Z(x^i)^T(\tilde{x}^{i+1} - x^i) \qquad (5.18)$$

$$\text{subject to } \tilde{x}^{i+1} \in \Omega, \text{ and}$$

$$\|y^{i+1}\| = \|\tilde{x}^{i+1} - x^i\| < \infty .$$

Avriel (1976) has pointed out that $\|y^{i+1}\| < \infty$ can be replaced by a set of conditions:

$$-U_1 I \leq y^{i+1} \leq U_2 I \text{ , } I = \text{identity matrix ;} \qquad (5.19)$$

that is, each component of y^{i+1} is bounded by positive scalars U_1 and U_2, which of course implies that $\|y^{i+1}\|$ is bounded. Using (5.19) and dropping the constant terms in the objective function, the LP becomes

$$\text{minimize } \nabla Z(x^i)\ \tilde{x}^{i+1} \qquad (5.20)$$

$$\text{subject to } \tilde{x}^{i+1} \in \Omega$$

$$x^i - U_1 I \leq \tilde{x}^{i+1} \leq x^i + U_2 I .$$

Therefore, the feasible direction vector is formed by setting $y^{i+1} = \tilde{x}^{i+1} - x^i$.

The next step in the feasible direction algorithm is to select a step length along the direction y^{i+1}. This step length, α^{i+1} must be selected by the following criterion (p. 447 of Avriel, 1976):

$$\alpha^{i+1} = \min \{\alpha_1^{i+1}, \alpha_2^{i+1}\} \qquad (5.21)$$

where

$$\alpha_1^{i+1} = \max\{\alpha\colon (y^{i+1})^T\ \nabla Z(x^i + \alpha y^{i+1}) \leq 0\} \qquad (5.22)$$

$$\alpha_2^{i+1} = \max\{\alpha\colon (x^i + \alpha y^{i+1}) \in \Omega\} . \qquad (5.23)$$

The above criterion selects α^{i+1} such that the step remains feasible and the function along y^{i+1} is minimized. If Z is convex, then unidimensional search methods such as a binary search or a Fibonacci search automatically satisfy (5.21). However, if Z is nonconvex, then more complicated line search procedures, such as the Armijo rule described in Bertsekas (1982), must be used.

Therefore, even if Z is nonconvex, if the direction vector and step size are chosen correctly, the feasible direction algorithm will converge to a (local) minima of Z (see Avriel, 1976, or Luenberger, 1973).

In order to implement this feasible direction method on (5.14), let us look more closely at the LP phase (5.20). The LP which needs to be solved at each iteration is is:

$$\text{minimize } \tilde{Z} = \sum_{\ell \varepsilon L} [\Psi_\ell^i S_\ell - \Theta_\ell^i D_\ell] + \sum_{a \varepsilon A} \Phi \tilde{t}_a^i f_a + \sum_{b \varepsilon B} MC_b^i e_b \tag{5.24}$$

subject to: $x = (S; D; f; e) \; \varepsilon \; \Omega^i$,

where Ω^i is the union of Ω with the bounds in (5.20). The solution to this LP will be denoted as $\tilde{x}^{i+1}$.

Following the traffic assignment algorithm of LeBlanc, Morlok and Pierskalla (1975), using the relationship $e_b = \sum_{q \varepsilon Q} \lambda_{b,q} g_q$ the last term of (5.24) can be written as:

$$\sum_{v \varepsilon V} \sum_{q \varepsilon Q_v} MC_q^i g_q . \tag{5.25}$$

To minimize $\tilde{Z}$, all the flow τ_v between a carrier O-D pair $v \varepsilon V$ should be loaded onto the shortest path $\bar{q}_v$. Letting MC_v^* be the cost of this path, we can rewrite (5.25) as:

$$\sum_{v \varepsilon V} \sum_{q \varepsilon Q_v} MC_v^* g_q = \sum_{v \varepsilon V} MC_v^* \sum_{q \varepsilon Q_v} g_q = \sum_{v \varepsilon V} MC_v^* \tau_v . \tag{5.26}$$

Using the relationship $[\chi_{a,v}]$ between shipper arc flows and carrier O-D flows, the last three terms in (5.24) can be written as:

$$\sum_{a \in A} (\Phi\, t_a^i + \sum_{v \in V} MC_v^*)\, f_a \tag{5.27}$$

Using the relationship between shipper arc and path flows, (5.27) can be written as:

$$\sum_{w \in W} \sum_{p \in P_w} TC_p\, h_p \,. \tag{5.28}$$

To minimize $\tilde{Z}$, all the flow T_w between a shipper O-D pair $w \in W$ should be loaded into the shortest path $\bar{p}_w$. Letting TC_w^* be the cost of this path, (5.28) can be rewritten as:

$$\sum_{w \in W} \sum_{p \in P_v} TC_w^*\, h_p = \sum_{w \in W} TC_w^* \sum_{p \in P_v} h_p = \sum_{w \in W} TC_w^*\, T_w \,. \tag{5.29}$$

The LP (5.24) thus becomes

$$\text{minimize } Z = \sum_{\ell \in L} \lfloor \Psi_\ell^i\, S_\ell - \Theta_\ell^i\, D_\ell \rfloor + \sum_{w \in W} TC_w^*\, T_w \tag{5.30}$$

subject to:

$$S_\ell - D_\ell \sum_{i \in L} \sum_{(i,\ell) \in W} T_{(i,\ell)} - \sum_{j \in L} \sum_{(\ell,j) \in W} T_{(\ell,j)} = 0$$

$$\forall\, \ell \in L \,,$$

$$x^i - U_1 \le x \le x^i + U_2 \,,$$

$$x \ge 0.$$

If the constraint

$$\sum_{\ell \in L} D_\ell - \sum_{\ell \in L} S_\ell = 0 \tag{5.31}$$

is added to (5.30), then the reader can easily verify that this LP is a network linear program. Therefore, the network LP can be stated as:

$$\text{minimize } Z = \sum_{\ell \in L} \lfloor \Psi_\ell^i\, S_\ell - \Theta_\ell^i\, D_\ell \rfloor + \sum_{w \in W} TC_w^*\, T_w \tag{5.32}$$

subject to

$$S_\ell - D_\ell + \sum_{i\varepsilon L} \sum_{(i,\ell)\varepsilon W} - \sum_{j\varepsilon L} \sum_{(\ell,j)\varepsilon W} T_{(\ell,j)} = 0 \quad \forall\ \ell \ \varepsilon\ L\ ,$$

$$\sum_{\ell\varepsilon L} D_\ell - \sum_{\ell\varepsilon L} S_\ell = 0,$$

$$\max\{S_\ell^i - U_1,\ 0\} \le S_\ell \le S_\ell^i + U_2 \qquad \forall\ \ell\ \varepsilon\ L$$

$$\max\{D_\ell^i - U_1,\ 0\} \le D_\ell \le D_\ell^i + U_2 \qquad \forall\ \ell\ \varepsilon\ L$$

$$\max\{T_w^i - U_1,\ 0\} \le T_w \le T_w^i + U_2 \qquad \forall\ w\ \varepsilon\ W$$

The last three constraints state that the solution to this LP must be feasible ($x \ge 0$) and not stray too far from the previous iteration's solution ($x^i - U_1 \le x \le x^i + U_2$). The bounds ($U_1$, U_2) may be arbitrarily large numbers, or they may arise from practical considerations. For example, if the inverse demand function at node ℓ is of the form

$$\Theta_\ell(D) = a - bD_\ell \quad \text{for } D_\ell \le \frac{a}{b}$$

$$\text{undefined for } D_\ell > \frac{a}{b}$$

where a and b are positive scalars, then D_ℓ must be constrained to lie in the interval $\lfloor 0,\ a/b \rfloor$. Therefore, the bounds (U_1, U_2) may play an important role in the definition of the inverse supply and demand functions. Problem (5.32) can be solved very efficiently with a number of network algorithms, such as out-of-kilter and network simplex. The other variables $\tilde{f}^{k+1}$ and $\tilde{e}^{k+1}$ are then derived by loading first the shipper shortest paths with $\tilde{T}^{i+1}$, and then the carrier shortest paths with the resulting $\tilde{\tau}^{k+1}$. This yields the LP solution vector $\tilde{x}^{k+1}$.

Therefore, the feasible direction algorithm for the variational inequality subproblem in ALG2 will be called ALG3, and can be stated as:

<u>Step 0</u> Choose an initial iterate $x^o \varepsilon \Omega$. Set $i = 0$.

<u>Step 1</u> Solve for the shortest path $\bar{q}_v \ \forall \ v \ \varepsilon \ V$.

<u>Step 2</u> Solve for the shortest path $\bar{p}_w \ \forall \ w \ \varepsilon \ W$.

<u>Step 3</u> Form the LP (5.32) and solve it for $\tilde{x}^{i+1}$.

<u>Step 4</u> Perform a linesearch along the direction $y^{i+1} = \tilde{x}^{i+1} - x^i$ to find the step length α^{i+1} which satisfies the criterion (5.21) - (5.23).

<u>Step 5</u> Set $x^{i+1} = x^i + \alpha^{i+1} y^{i+1}$.

<u>Step 6</u> If

$$|S_\ell^{i+1} - S_\ell^i| \leq \varepsilon_9 \qquad \forall \ \ell \ \varepsilon \ L,$$

$$|D_\ell^{i+1} - D_\ell^i| \leq \varepsilon_{10} \qquad \forall \ \ell \ \varepsilon \ L,$$

$$|f_a^{i+1} - f_a^i| \leq \varepsilon_{11} \qquad \forall \ a \ \varepsilon \ A, \text{ and}$$

$$|e_b^{i+1} - e_b^i| \leq \varepsilon_{12} \qquad \forall \ b \ \varepsilon \ B,$$

then stop; x^{i+1} is a solution to (5.14), and thus is a solution to the subproblem in step 2 of ALG2. Else, set $i = i + 1$ and go to step 1.

To summarize the preceding discussion, the diagonalization algorithm, ALG2, can be stated as

<u>Step 0</u> Choose an initial iterate $x^o = (S^o; D^o; f^o; e^o) \ \varepsilon \ \Omega$. Set $k = 0$.

<u>Step 1</u> Solve the separable VIP/optimization problem by ALG3.

<u>Step 2</u> If

$$|S_\ell^{k+1} - S_\ell^k| \leq \varepsilon_5 \qquad \forall \ \ell \ \varepsilon \ L,$$

$$|D_\ell^{k+1} - D_\ell^k| \leq \varepsilon_6 \qquad \forall \ \ell \ \varepsilon \ L,$$

$$|f_a^{k+1} - f_a^k| \le \varepsilon_7 \qquad \forall \ a \in A,$$

$$|e_b^{k+1} - e_b^k| \le \varepsilon_8 \qquad \forall \ b \in B,$$

then stop; x^{k+1} is a solution to (5.9). Else, set $k = k + 1$ and return to step 1.

For monotonic functions, the convergence of ALG2 and ALG3 is assured locally, and if the functions are strictly monotone, then global convergence is guaranteed. For nonmonotonic functions, the convergence of ALG3 to a solution can still be assured by properly defining the step-size and direction of descent, but the convergence of ALG2 is difficult to insure in general. However, if at a solution the functions are strictly monotone, then local convergence can be assured.

The convergence rate of ALG3, the Frank-Wolfe based algorithm, is sublinear (see Dunn, 1979, and Canon and Cullum, 1968), therefore the speed of convergence for ALG2 will be slow. However, the computer storage requirements of this algorithm are small compared with ALG1 due to the fact that the path sets do not need to be stored from iteration to iteration. Therefore, although it has quadratic convergence, ALG1 will take much more storage than ALG2 to implement. We shall discuss alternatives to this ALG2/ALG3 approach in Section 5.4.

The next section describes some numerical experiments with these two algorithms, and a hybrid of the two which we will call ALG4.

5.3 Numerical Examples

In this section, a series of test examples is solved to study the properties of the algorithms presented in this chapter. A hybrid algorithm, called ALG4, will also be used on these examples. In this algorithm, ALG2 will be used on a problem regardless of the assumed pricing so that a good initial set of paths and vector x^o can be generated for ALG1. Thus, ALG4 uses in sequence ALG2 and ALG1 to solve a problem. In summary, the algorithms which will be used on these test problems are:

ALG1 - iterative complementarity algorithm,

ALG2 - diagonalization/relaxation algorithm,

ALG3 - a Frank-Wolfe-type feasible direction algorithm for the subproblems generated by ALG2,

ALG4 - a hybrid algorithm, using ALG2 and ALG1 in sequence.

In the implementation of ALG1 and ALG3 for the examples in this section, the computer code by Tomlin (1976a, b) is used to solve the linear complementarity problems via Lemke's algorithm, the out-of-kilter algorithm is used to solve the network LP in ALG3, and Dijkstra's algorithm is used to find shortest paths in the networks. Also, the stopping tolerances in the algorithms have been slightly altered. Instead of the stopping criterion being in absolute units $|x_i^{i+1} - x_i^k|$, the stopping criterion used for the examples in this section are in percentage terms:

$$|x_i^{k+1} - x_i^k|/x_k^k \le \varepsilon \ . \tag{5.33}$$

Finally, all the problems we will consider have monotonic functions, and thus a bisection linesearch will be used.

Let us define the functional forms used in these examples. Let

M = the set of commodities which are considered in a problem,

m,n = indices of commodities,

a = (i,j); that is, an arc is denoted by its tail node i and head node j.

The marginal cost functions for each carrier arc and commodity take the following form, where XH_b^m, XI_b^m, and $XJ_b^{m,n}$ are coefficients:

$$MC_b^m(e) = XH_b^m + 2*XI_b^m e_b^m + 1.5*XJ_b^{m,m} (e_b^m)^2$$

$$+ \sum_{n \varepsilon M} XJ_b^{m,n} e_b^m e_b^n \quad \forall\ b \ \varepsilon\ B,\ m\ \varepsilon\ M\ . \qquad (5.34)$$

The time delay functions are exogenously aggregated; that is, the time delay functions are defined directly on the shipper arcs. The form of these functions is, where XK_a^m, XL_a^m and $XM_a^{m,n}$ are coefficients:

$$t_a^m = XK_a^m + XL_a^m (f_a^m)^4 + 0.5* \sum_{n \varepsilon M} XM_a^{m,n} f_a^m f_a^n \quad \forall\ a\ \varepsilon\ A,\ m\ \varepsilon\ M\ . \qquad (5.35)$$

The supply and demand functions are given by

$$S_\ell^m = XA_\ell^m + XB_\ell^m \pi_\ell^m + 0.5* \sum_{n \varepsilon M} XC_\ell^{m,n} \pi_\ell^m \pi_\ell^n \quad \forall\ \ell\ \varepsilon\ L,\ m\ \varepsilon\ M\ . \qquad (5.36)$$

$$D_\ell^m = XE_\ell^m + XF_\ell^m \pi_\ell^m + 0.5* \sum_{n \varepsilon M} XG_\ell^{m,n} \pi_\ell^m \pi_\ell^n \quad \forall\ \ell\ \varepsilon\ L,\ m\ \varepsilon\ M\ . \qquad (5.37)$$

where XA_ℓ^m, XB_ℓ^m, $XC_\ell^{m,n}$, XE_ℓ^m, XF_ℓ^m and $XG_\ell^{m,n}$ are coefficients. The reader can easily see that if there is only a single commodity under study, then all functions used are separable. If there are multiple commodities, the degree of asymmetry of the Jacobian is determined by the values of $XJ_b^{m,n}$, $XM_a^{m,n}$, $XG_\ell^{m,n}$, and $XC_\ell^{m,n}$.

To monitor how exact each algorithm is, that is, how close the solution of each algorithm comes to the true equilibrium solution, the following error measures will be used:

$$\text{ERROR}_1 = 100*\max\{\frac{|\Theta^m_\ell - \Psi^m_\ell|}{0.5(\Theta^m_\ell + \Psi^m_\ell)} \quad \forall\ \ell \in L,\ m \in M\},$$

$$\text{ERROR}_2 = 100*\max\{\frac{|\pi^m_i + TC^{*m}_{i,j} - \pi^m_j|}{\pi^m_i + TC^{*m}_{i,j}} \quad \forall (i,j) \in W,\ m \in M\text{: } T^m_{ij} > 0\},$$

$$\text{ERROR}_3 = 100*\max\{\frac{|TC^{*m}_{i,j} - TC_p|}{TC^{*m}_{i,j}} \quad \forall (i,j) \in W,\ p \in P_{(i,j)}\text{: } h_p > 0\}$$

$$\text{ERROR}_4 = 100*\max\{\frac{|MC^*_v - MC_q|}{MC^*_v} \quad \forall\ v \in V,\ q \in Q_v\text{: } g_q > 0\}$$

The first error term measures how far, in percentage terms, the inverse supply and demand functions are from equaling each other. When discussing ALG1, this error measure is identically zero due to the fact that prices are dealt with directly in the complementarity formulation. The remaining error terms measure how far, in percentage terms, the equilibrium conditions are from being met exactly.

The networks used in the first set of examples are shown in Figure 5.1, where each arc is bi-directional. A single commodity problem, Example 1A and a two commodity problem, Example 1B, are defined using these networks. Let us begin with the single commodity example.

The data used in Example 1A is listed in Table 5.1. The first problem to be solved using this data is the case of marginal cost pricing ($R_V = MC_V^{*m}\ \forall\ v \in V$). First, the complementarity algorithm, ALG1, was run, and the results of this run are listed in Table 5.2. The algorithm took 3.14 CPU seconds, and yielded very small errors in the equilibrium conditions. Next, the feasible direction/diagonalization algorithm, ALG2, was run. As the results in Table 5.3 show, ALG2 took slightly longer, 3.38 seconds, and yielded much greater errors. Lastly, ALG2 was run

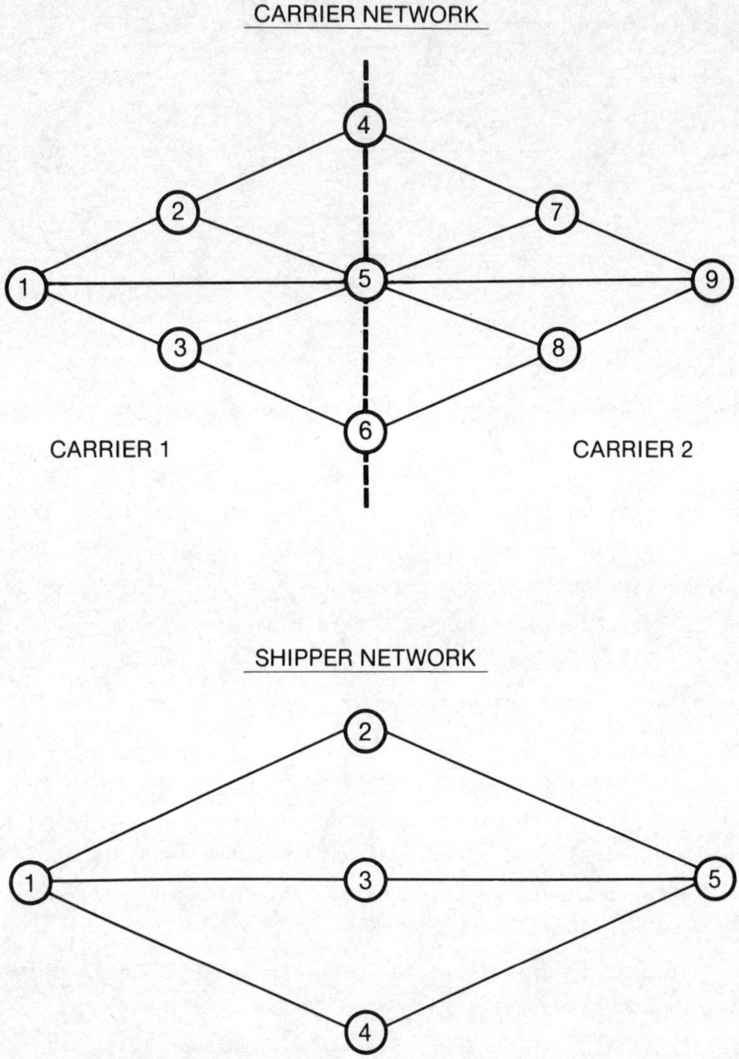

Figure 5.1: Test Networks for Example 1

with loose stopping tolerances, and then the results of this run were passed to ALG1 to complete the calculations (ALG4). Table 5.4 summarizes the results of ALG4. ALG2 took less iterations, in terms of the ALG3 iterations, than when it was run to a complete solution, 9 vs. 13. Also, ALG1 took less iterations with this approach, 9 vs. 13.

The reader should note by looking at Tables 5.3 and 5.4 that the Frank-Wolfe-type path enumeration scheme and the partial path enumeration scheme used in ALG1 yield the exact same path sets. Thus, the path enumeration scheme used in conjunction with ALG1 does indeed act in the same way as that of Frank-Wolfe decomposition.

Also, the reader can see the difference between quadratic convergence (ALG1) and sublinear convergence (ALG3). ALG1 yields a more precise answer in less time than ALG2, which uses a Frank-Wolfe sub-algorithm (ALG3). The 'tailing' phenomenon characteristic of Frank-Wolfe-type algorithms is evident in the high error values.

The hybrid algorithm, ALG4, shows no distinct total advantage over either ALG1 or ALG2 in this small example. However, in Example 1B, the hybrid approach will prove useful.

To show that the model can be solved with non-marginal cost pricing assumptions, Tables 5.5 and 5.6 show the results of applying ALG4 to problems with

$$R_v = 2.0 * MC_v^*$$

and

$$R_v = 10.0 + 0.5 * MC_v^*$$

respectively. In both cases, marginal cost pricing was first assumed, and ALG2 was used with loose stopping tolerances to form an initial solution for ALG1. Then, ALG1 was used to calculate the solution with the correct rate specification. Again, very small error values are achieved by ALG1 in relatively little time.

TABLE 5.1

DATA FOR EXAMPLE 1A

No. of carriers = 2

No. of carrier arcs = 28

No. of carrier nodes = 9

No. of shipper arcs = 12

No. of shipper nodes = 5

Carrier Network Data

Arc (i,j)	$XH^{1}_{i,j}$	$XI^{1}_{i,j}$	$XJ^{1,1}_{i,j}$
1,2	4.0	0.1	0.01
1,3	3.0	0.2	0.02
1,5	2.0	0.2	0.03
2,1	1.0	0.1	0.04
2,4	4.0	0.1	0.01
2,5	3.0	0.2	0.04
3,1	2.0	0.2	0.03
3,5	1.0	0.1	0.04
3,6	4.0	0.1	0.01
4,2	3.0	0.2	0.02
4,7	2.0	0.2	0.03
5,1	1.0	0.1	0.04
5,2	4.0	0.1	0.01
5,3	3.0	0.2	0.02
5,7	2.0	0.2	0.03
5,8	1.0	0.1	0.04
5,9	4.0	0.1	0.01

Arc (i,j)	$XH^{1}_{i,j}$	$XI^{1}_{i,j}$	$XJ^{1,1}_{i,j}$
6,3	3.0	0.2	0.02
6,8	2.0	0.2	0.02
7,4	1.0	0.1	0.04
7,5	4.0	0.1	0.01
7,9	3.0	0.2	0.02
8,5	2.0	0.2	0.03
8,6	1.0	0.1	0.04
8,9	4.0	0.1	0.01
9,5	3.0	0.2	0.02
9,7	2.0	0.2	0.03
9,8	1.0	0.1	0.04

Shipper Network Data

Arc (i,j)	$XK^{1}_{i,j}$	$XL^{1}_{i,j}$	$XM^{1,1}_{i,j}$
1,2	1.0	0.001	0.01
1,3	2.0	0.002	0.02
1,4	1.0	0.002	0.09
2,1	2.0	0.001	0.01
2,5	3.0	0.002	0.02
3,1	1.0	0.002	0.03
3,5	1.0	0.001	0.01
4,1	2.0	0.004	0.01
4,5	1.0	0.001	0.01
5,2	3.0	0.001	0.02
5,3	6.0	0.002	0.05
5,4	1.0	0.005	0.03

Node ℓ	XA^{1}_{ℓ}	XB^{1}_{ℓ}	$XC^{1,1}_{\ell}$	XE^{1}_{ℓ}	XF^{1}_{ℓ}	$XG^{1,1}_{\ell}$
1	0.0	0.1	0.1	400.0	-2.0	-9.0
5	0.0	0.01	0.05	500.0	-1.0	-0.1

$\Phi = 1.0$

TABLE 5.2

RESULTS OF APPLYING ALG1, THE ITERATIVE COMPLEMENTARITY ALGORITHM, TO EXAMPLE 1A, MARGINAL COST PRICING

$\varepsilon_1 = \varepsilon_2 = \varepsilon_3 = \varepsilon_4 = 1\%$

No. of iterations = 13

CPU time, sec* = 3.14

No. of shortest path calls = 154

Shipper Network Results

Node ℓ	π^1_ℓ	S^1_ℓ	D^1_ℓ	$S^1_\ell - D^1_\ell$
1	27.624	40.917	1.358	39.560
5	71.908	129.990	169.550	-39.560

O-D Pair (i,j)	$T^1_{(i,j)}$	$TC^{*1}_{(i,j)}$	Path p	Node Sequence Comprising p	h_p	TC_p
1-5	39.560	44.280	1	1-4-5	8.029	44.280
			2	1-3-5	21.724	44.295
			3	1-2-5	9.807	44.283
5-1	0.0	3.000	1	5-4-1	0.0	3.000

Carrier Network Results

O-D Pair (i,j)	$\tau^1_{(i,j)}$	$MC^{*1}_{(i,j)}$	Path q	Node Sequence Comprising q	g_q	MC_q
1-4	9.807	17.249	1	1-2-4	9.807	17.250
1-5	21.724	15.161	1	1-5	13.225	15.161
			2	1-2-5	4.362	15.161
			3	1-3-5	4.137	15.161

TABLE 5.2 (continued)

O-D Pair (i,j)	$\tau^{1}_{(i,j)}$	$MC^{*1}_{(i,j)}$	Path q	Node Sequence Comprising q	g_q	MC_q
1-6	8.29	18.879	1	1-3-6	8.029	18.879
4-9	9.807	20.119	1	4-7-9	9.807	20.119
	21.724	11.897	1	5-9	17.224	11.897
			2	5-7-9	0.062	11.897
			3	5-8-9	4.438	11.897
6-9	8.029	16.935	1	6-8-9	8.029	16.935

$ERROR_1$ = 0%

$ERROR_2$ = 0.00%

$ERROR_3$ = 0.03%

$ERROR_4$ = 0.02%

*on an IBM 4031 using Fortran H, optimization level 3.

TABLE 5.3

RESULTS OF APPLYING ALG2, THE FRANK-WOLFE BASED FEASIBLE DIRECTION ALGORITHM, TO EXAMPLE 1A, MARGINAL COST PRICING

$\varepsilon_5 = \varepsilon_6 = \varepsilon_7 = \varepsilon_8 = 5\%$

$\varepsilon_9 = \varepsilon_{10} = \varepsilon_{11} = \varepsilon_{12} = 2\%$

No. of iterations for ALG2 = 2

No. of iterations for ALG3 = 9

CPU time, sec* = 3.38

No. of shortest path calls = 114

Shipper Network Results

Node ℓ	Ψ^1_ℓ	Θ^1_ℓ	S^1_ℓ	D^1_ℓ	$S^1_\ell - D^1_\ell$
1	30.659	27.205	50.063	12.530	37.533
5	72.002	72.114	130.328	167.862	-37.533

O-D Pair (i,j)	$T^1_{(i,j)}$	$TC^{*1}_{(i,j)}$	Path p	Node Sequence Compri-sing p	h_p	TC_p
1-5	37.533	40.889	1	1-4-5	7.911	42.918
			2	1-3-5	20.841	41.276
			3	1-2-5	8.782	40.889
5-1	0.0	3.000	1	5-4-1	0.0	3.000

TABLE 5.3 (continued)

Carrier Network Results

O-D Pair (i,j)	$\tau^{1}_{(i,j)}$	$MC^{*1}_{(i,j)}$	Path q	Node Sequence Compri-sing q	g_q	MC_q
1-4	8.782	15.976	1	1-2-4	8.782	15.976
1-5	20.841	14.203	1	1-5	12.984	14.780
			2	1-2-5	4.094	14.359
			3	1-3-5	3.763	14.203
1-6	7.911	18.278	1	1-3-6	7.911	18.278
4-9	8.782	18.574	1	4-7-9	8.782	18.574
5-9	20.841	10.980	1	5-9	16.276	11.229
			2	5-7-9	0.803	11.946
			3	5-8-9	3.763	10.980
6-9	7.911	16.359	1	6-8-9	7.911	16.359

$ERROR_1$ = 11.94%

$ERROR_2$ = 3.20%

$ERROR_3$ = 4.96%

$ERROR_4$ = 8.73%

*on an IBM 4031 using Fortran H, optimization level 3.

TABLE 5.4

RESULTS OF APPLYING ALG4, THE HYBRID ALGORITHM, TO EXAMPLE 1A, MARGINAL COST PRICING

$\varepsilon_1 = \varepsilon_2 = \varepsilon_3 = \varepsilon_4 = 1\%$

$\varepsilon_5 = \varepsilon_6 = \varepsilon_7 = \varepsilon_8 = 20\%$

$\varepsilon_9 = \varepsilon_{10} = \varepsilon_{11} = \varepsilon_{12} = 15\%$

Results of ALG2

No. of iterations for ALG2	=	2
No. of iterations for ALG3	=	9
CPU time, sec*	=	3.58
No. of shortest path calls	=	96

Results of ALG1

No. of iterations	=	9
CPU time, sec*	=	1.89
No. of shortest path calls	=	106

Totals

Total CPU time, sec*	=	5.47
Total no. of shortest path calls	=	202

(results are the same as in Table 5.2)

*on an IBM 4031 using Fortran H, optimization level 3.

The next example, Example 1B, is double the size of the preceding example due to the presence of two commodities. The data for this example is listed in Table 5.7. As the reader can see, asymmetries exist in the functions defined by this dataset. Tables 5.8 and 5.9 present the results of solving the marginal cost pricing version of this example by ALG2 and the hybrid algorithm, ALG4. The 'tailing' of ALG2 is even more pronounced in this example, reflected in the high error values. In less than half the CPU time, 6.28 seconds vs. 14.65 seconds, the hybrid algorithm has solved the model to the point where the errors in the equilibrium conditions are essentially zero.

Tables 5.10 and 5.11 present a summary of the results of examples with

$$R_v^1 = 1.8 * MC_v^*$$

$$R_v^2 = 1.6 * MC_v^*$$

and

$$R_v^1 = 8.0 + 0.9 * MC_v^*$$

$$R_v^2 = 0.6 + 0.6 * MC_v^*$$

respectively. Again, ALG4 proves to be able to solve these problems accurately and efficiently.

An example with larger networks, Example 2, has also been solved assuming the rates equal marginal costs. The carrier network and shipper network for this example are shown in Figures 5.2 and 5.3 respectively. A brief description of the data used in this example is given in Table 5.12. ALG2 was applied to this problem and the results are summarized in Table 5.13. Table 5.14 presents the results of applying ALG4, the hybrid algorithm, to this example. Again, the 'tailing' of ALG3 is evident by the high error values in Table 5.13.

TABLE 5.5

RESULTS OF APPLYING ALG4, THE HYBRID ALGORITHM, TO EXAMPLE 1A, RATE = 2.0* MARGINAL COST

$\varepsilon_1 = \varepsilon_2 = \varepsilon_3 = \varepsilon_4 = 1\%$

$\varepsilon_5 = \varepsilon_6 = \varepsilon_7 = \varepsilon_8 = 20\%$

$\varepsilon_9 = \varepsilon_{10} = \varepsilon_{11} = \varepsilon_{12} = 15\%$

Results of ALG2

No. of iterations for ALG2	=	2
No. of iterations for ALG3	=	8
CPU time, sec*	=	3.56
No. of shortest path calls	=	96

Results of ALG1

No. of iterations	=	4
CPU time, sec*	=	0.88
No. of shortest path calls	=	46

Totals

Total CPU time, sec*	=	4.44
Total no. of shortest path calls	=	142

TABLE 5.5 (continued)

Shipper Network Results

Node ℓ	π_ℓ^1	S_ℓ^1	D_ℓ^1	$S_\ell^1 - D_\ell^1$
1	27.050	39.291	16.629	22.662
5	73.328	135.159	157.820	-22.661

O-D Pair (i,j)	$T^1_{(i,j)}$	$TC^{*1}_{(i,j)}$	Path p	Node Sequence Comprising p	h_p	TC_p
1-5	22.662	46.148	1	1-4-5	3.165	46.313
			2	1-3-5	15.512	46.148
			3	1-2-5	3.985	46.289
5-1	0.0	3.000	1	5-4-1	0.0	3.000

Carrier Network Results

O-D Pair (i,j)	$\tau^1_{(i,j)}$	$MC^{*1}_{(i,j)}$	Path q	Node Sequence Comprising q	g_q	MC_q
1-4	3.985	10.853	1	1-2-4	3.985	10.853
1-5	15.512	9.831	1	1-5	9.497	9.831
			2	1-2-5	2.223	9.831
			3	1-3-5	3.793	9.831
1-6	3.165	12.019	1	1-3-6	3.165	12.019
4-9	3.985	8.152	1	5-9	11.289	8.151
			2	5-7-9	0.988	18.152
			3	5-8-9	3.235	8.152
6-9	3.165	9.625	1	6-8-9	3.165	9.625

$ERROR_1$ = 0% $ERROR_2$ = 0.00% $ERROR_3$ = 0.36% $ERROR_4$ = 0.27%

*on an IBM 4031 using Fortran H, optimization level 3.

TABLE 5.6

RESULTS OF APPLYING ALG4, THE HYBRID ALGORITHM, TO EXAMPLE 1A, RATE = 10.0 + 0.5* MARGINAL COST

$\varepsilon_1 = \varepsilon_2 = \varepsilon_3 = \varepsilon_4 = 1\%$

$\varepsilon_5 = \varepsilon_6 = \varepsilon_7 = \varepsilon_8 = 20\%$

$\varepsilon_9 = \varepsilon_{10} = \varepsilon_{11} = \varepsilon_{12} = 15\%$

Results of ALG2

No. of iterations for ALG2	=	2
No. of iterations for ALG3	=	8
CPU time, sec*	=	3.56
No. of shortest path calls	=	96

Results of ALG1

No. of iterations	=	5
CPU time, sec*	=	1.11
No. of shortest path calls	=	58

Totals

Total CPU time, sec*	=	4.67
Total no. of shortest path calls	=	154

TABLE 5.6 (continued)

Shipper Network Results

Node ℓ	π_ℓ^1	S_ℓ^1	D_ℓ^1	$S_\ell^1 - D_\ell^1$
1	27.488	40.529	4.999	35.530
5	72.249	131.222	166.752	-35.530

O-D Pair (i,j)	$T_{(i,j)}^1$	$TC_{(i,j)}^{*1}$	Path p	Node Sequence Comprising p	h_p	TC_p
1-5	35.530	44.662	1	1-4-5	7.810	44.662
			2	1-3-5	18.071	44.977
			3	1-2-5	9.649	44.729
5-1	0.0	23.000	1	5-4-1	0.0	23.000

Carrier Network Results

O-D Pair (i,j)	$\tau_{(i,j)}^1$	$MC_{(i,j)}^{*1}$	Path q	Node Sequence Comprising q	g_q	MC_q
1-4	9.649	16.211	1	1-2-4	0.649	16.211
1-5	18.071	13.355	1	1-5	11.995	13.355
			2	1-2-5	2.890	13.355
			3	1-3-5	3.186	13.356
1-6	7.810	17.493	1	1-3-6	7.819	17.493
4-9	9.649	19.637	1	4-7-9	9.649	19.637
5-9	18.071	10.369	1	5-9	14.861	10.369
			2	5-7-9	0.0	11.624
			3	5-8-9	3.210	10.371
6-9	7.810	15.827	1	6-8-9	7.810	15.827

$ERROR_1 = 0\%$ $ERROR_2 = 0.00\%$ $ERROR_3 = 0.71\%$ $ERROR_4 = 0.83\%$

*on an IBM 4031 using Fortran H, optimization level 3.

TABLE 5.7

DATA FOR EXAMPLE 1B

No. of carriers	=	2
No. of carrier arcs	=	28
No. of carrier nodes	=	9
No. of shipper arcs	=	12
No. of shipper nodes	=	5
No. of commodities	=	2

Carrier Network Data

Arc (i,j)	Commodity m	$XH^m_{i,j}$	$XI^m_{i,j}$	$XJ^{m,1}_{i,j}$	$XJ^{m,2}_{i,j}$
1,2	1	4.0	0.1	0.010	0.002
	2	2.0	0.2	0.002	0.010
1,3	1	3.0	0.2	0.020	0.003
	2	1.0	0.3	0.002	0.030
1,5	1	2.0	0.2	0.030	0.001
	2	2.0	0.1	0.002	0.030
2,1	1	1.0	0.1	0.040	0.007
	2	3.0	0.2	0.004	0.010
2,4	1	4.0	0.1	0.010	0.002
	2	2.0	0.2	0.002	0.016
2,5	1	3.0	0.2	0.020	0.003
	2	1.0	0.3	0.002	0.030
3,1	1	2.0	0.2	0.030	0.001
	2	2.0	0.1	0.002	0.030
3,5	1	1.0	0.1	0.040	0.007
	2	3.0	0.2	0.004	0.010
3,6	1	4.0	0.1	0.010	0.002
	2	2.0	0.2	0.002	0.010
4,2	1	3.0	0.2	0.020	0.003
	2	1.0	0.3	0.002	0.030
4,7	1	2.0	0.2	0.030	0.001
	2	2.0	0.1	0.002	0.030
5,1	1	1.0	0.2	0.040	0.007
	2	3.0	0.2	0.004	0.010
5,2	1	4.0	0.1	0.010	0.002
	2	2.0	0.2	0.002	0.010

TABLE 5.7 (continued)

Arc (i,j)	Commodity m	$XH^{m}_{i,j}$	$XI^{m}_{i,j}$	$XJ^{m,1}_{i,j}$	$XJ^{m,2}_{i,j}$
5,3	1	3.0	0.2	0.002	0.003
	2	1.0	0.3	0.002	0.030
5,7	1	2.0	0.2	0.030	0.001
	2	2.0	0.1	0.002	0.030
5,8	1	1.0	0.1	0.040	0.007
	2	3.0	0.2	0.004	0.010
5,9	1	4.0	0.1	0.010	0.002
	2	2.0	0.2	0.002	0.010
6,3	1	3.0	0.2	0.020	0.010
	2	1.0	0.2	0.002	0.030
6,8	1	2.0	0.2	0.030	0.001
	2	2.0	0.1	0.002	0.030
7,4	1	1.0	0.1	0.040	0.007
	2	3.0	0.2	0.004	0.010
7,5	1	4.0	0.1	0.010	0.002
	2	2.0	0.2	0.002	0.010
7,9	1	3.0	0.2	0.020	0.003
	2	1.0	0.3	0.002	0.030
8,5	1	2.0	0.2	0.030	0.001
	2	2.0	0.1	0.002	0.030
8,6	1	1.0	0.1	0.040	0.007
	2	3.0	0.2	0.004	0.020
8,9	1	4.0	0.1	0.010	0.002
	2	2.0	0.2	0.002	0.010
9,5	1	3.0	0.2	0.020	0.003
	2	1.0	0.3	0.002	0.030
9,7	1	1.0	0.2	0.030	0.001
	2	2.0	0.1	0.002	0.036
9,8	1	1.0	0.1	0.040	0.007
	2	3.0	0.2	0.004	0.010

Shipper Network Data

Arc (i,j)	Commodity m	$XH^{m}_{i,j}$	$XI^{m}_{i,j}$	$XJ^{m,1}_{i,j}$	$XJ^{m,2}_{i,j}$
1,2	1	1.0	0.1	0.010	0.004
	2	2.0	0.2	0.001	0.040
1,3	1	2.0	0.2	0.030	0.006
	2	1.0	0.1	0.003	0.060
1,4	1	3.0	0.2	0.020	0.002
	2	2.0	0.1	0.002	0.020

TABLE 5.7 (continued)

Arc (i,j)	Commodity m	$XH^m_{i,j}$	$XI^m_{i,j}$	$XJ^{m,1}_{i,j}$	$XJ^{m,2}_{i,j}$
2,1	1	1.0	0.3	0.010	0.006
	2	2.0	0.1	0.001	0.006
2,5	1	2.0	0.3	0.040	0.001
	2	1.0	0.2	0.004	0.010
3,1	1	1.0	0.2	0.020	0.002
	2	2.0	0.3	0.002	0.020
3,5	1	3.0	0.1	0.040	0.009
	2	5.0	0.3	0.004	0.090
4,1	1	1.0	0.3	0.020	0.003
	2	1.0	0.2	0.020	0.003
4,5	1	2.0	0.6	0.010	0.001
	2	5.0	0.8	0.001	0.010
5,2	1	1.0	0.9	0.010	0.006
	2	3.0	0.9	0.001	0.060
5,3	1	4.0	0.1	0.050	0.003
	2	1.0	0.1	0.010	0.002
5,4	1	6.0	0.3	0.001	0.020
	2	1.0	0.3	0.002	0.020

Node ℓ	Commodity m	XA^m_ℓ	XB^m_ℓ	$XC^{m,1}_\ell$	$XC^{m,2}_\ell$	XE^m_ℓ	XF^m_ℓ	$XG^{m,1}_\ell$
1	1	0.0	0.10	0.010	0.002	400.0	-3.0	-0.010
	2	0.0	0.01	0.008	0.040	800.0	-1.0	-0.001
5	1	0.0	0.01	0.050	0.001	500.0	-0.5	-0.010
	2	0.0	0.01	0.007	0.020	300.0	-4.0	-0.001

Node ℓ	Commodity m	$XG^{m,2}_\ell$
1	1	-0.003
	2	-0.030
5	1	-0.002
	2	-0.009

$\Phi = 1.0$

TABLE 5.8

RESULTS OF APPLYING ALG2, THE ITERATIVE COMPLEMENTARITY ALGORITHM, TO EXAMPLE 1B, MARGINAL COST PRICING

$\varepsilon_5 = \varepsilon_6 = \varepsilon_7 = \varepsilon_8 = 5\%$

$\varepsilon_9 = \varepsilon_{10} = \varepsilon_{11} = \varepsilon_{12} = 2\%$

No. of iterations for ALG2 = 2

No. of iterations for ALG3 = 142

CPU time, sec* = 14.65

No. of shortest path calls = 2982

Shipper Network Results

Commodity m	Node ℓ	Ψ_ℓ^m	Θ_ℓ^m	S_ℓ^m	D_ℓ^m	$S_\ell^m - D_\ell^m$
1	1	97.146	97.986	56.902	28.847	28.055
	5	108.914	138.407	297.645	325.654	-28.009
2	1	121.743	142.922	297.645	343.198	-45.553
	5	68.901	67.076	48.163	2.579	45.584

Commodity m	O-D Pair (i,j)	$T^m_{(i,j)}$	$TC^{*m}_{(i,j)}$	Path p	Node Sequence Comprising p	h_p	TC_p
1	1-5	28.037	39.590	1	1-4-5	6.510	39.681
				2	1-3-5	13.226	39.590
				3	1-2-5	8.301	40.433
	5-1	0.0	9.000	1	5-2-1	0.0	9.000
2	1-5	0.0	10.000	1	1-2-5	0.0	10.000
	5-1	48.578	72.326	1	5-3-1	22.475	72.326
				2	5-2-1	10.234	73.061
				3	5-4-1	12.869	72.404

TABLE 5.8 (continued)

Carrier Network Results

Commodity m	O-D Pair (i,j)	$\tau^m_{(i,j)}$	$MC^{*m}_{(i,j)}$	Path q	Node Sequence Comprising q	g_q	MC_q
1	1-4	9.301	13.859	1	1-2-4	8.301	13.859
	1-5	13.226	10.318	1	1-5	9.859	10.318
				2	1-3-5	2.350	10.712
				3	1-2-5	1.016	10.603
	1-6	6.510	14.837	1	1-3-6	6.510	14.837
	4-9	8.301	16.809	1	4-7-9	8.301	16.809
	5-9	13.226	8.060	1	5-9	11.085	8.387
				2	5-7-9	0.0	10.387
				3	5-8-9	2.141	8.560
	6-9	6.510	13.364	1	6-8-9	6.510	13.364
2	4-1	10.234	23.433	1	4-2-11	10.234	23.433
	5-1	22.475	14.961	1	5-3-1	1.487	15.086
				2	5-1	17.175	14.961
				3	5-2-1	3.813	15.322
	6-1	12.869	29.269	1	6-3-1	12.149	29.269
				2	6-3-5-1	0.720	34.437
	9-4	10.234	22.824	1	9-7-4	10.234	22.824
	9-5	22.824	17.873	1	9-5	13.549	17.873
				2	9-7-5	4.131	18.068
				3	9-8-5	4.795	18.495
	9-6	12.869	25.133	1	9-8-6	12.605	25.133
				2	9-5-8-6	0.264	31.613

$ERROR_1$ = 23.85%
$ERROR_2$ = 9.84%
$ERROR_3$ = 2.13%
$ERROR_4$ = 25.78%

*on an IBM 4031 using Fortran H, optimization level 3.

TABLE 5.9

RESULTS OF APPLYING ALG4, THE HYBRID ALGORITHM, TO EXAMPLE 1B, MARGINAL COST PRICING

$\varepsilon_1 = \varepsilon_2 = \varepsilon_3 = \varepsilon_4 = 5\%$

$\varepsilon_5 = \varepsilon_6 = \varepsilon_7 = \varepsilon_8 = 20\%$

$\varepsilon_9 = \varepsilon_{10} = \varepsilon_{11} = \varepsilon_{12} = 15\%$

Results of ALG2

No. of iterations for ALG2	=	5
No. of iterations for ALG3	=	32
CPU time, sec*	=	3.74
No. of shortest path calls	=	672

Results of ALG1

No. of iterations	=	5
CPU time, sec*	=	2.54
No. of shortest path calls	=	116

Totals

Total CPU time, sec*	=	6.28
Total no. of shortest path calls	=	788

TABLE 5.9 (continued)

Shipper Network Results

Commodity m	Node ℓ	π_ℓ^m	S_ℓ^m	D_ℓ^m	$S_\ell^m - D_\ell^m$
1	1	97.541	566.255	53.520	12.736
	5	117.760	351.592	364.328	-12.736
2	1	128.105	377.946	419.674	-41.728
	5	63.296	66.786	25.059	41.728

Commodity m	O-D Pair (i,j)	$T^m_{(i,j)}$	$TC^{*m}_{(i,j)}$	Path p	Node Sequence Comprising p	h_p	TC_p
1	1-5	12.736	23.204	1	1-4-5	1.850	23.204
				2	1-3-5	7.490	23.306
				3	1-2-5	3.396	23.204
	5-1	0.0	2.000	1	5-2-1	0.0	2.000
2	1-5	0.0	3.000	1	1-2-5	0.0	10.000
	5-1	41.727	64.777	1	5-3-1	20.702	64.932
				2	5-2-1	9.279	64.777
				3	5-4-1	11.747	64.785

TABLE 5.9 (continued)

Carrier Network Results

Commodity m	O-D Pair (i,j)	$\tau^{m}_{(i,j)}$	$MC^{*m}_{(i,j)}$	Path q	Node Sequence Comprising q	g_q	MC_q
1	1-4	3.396	9.700	1	1-2-4	3.396	9.700
	1-5	7.490	6.096	1	1-5	6.055	6.204
				2	1-3-5	1.434	6.098
				3	1-2-5	0.0	7.850
	1-6	1.850	9.066	1	1-3-6	1.850	9.066
	4-9	3.396	8.573	1	4-7-9	3.396	8.573
	5-9	7.490	5.953	1	5-9	6.518	5.953
				2	5-7-9	0.0	6.700
				3	5-8-9	0.972	5.954
	6-9	1.850	7.572	1	6-8-9	1.850	7.572
2	4-1	9.279	20.705	1	4-2-1	9.279	20.705
	5-1	20.702	13.691	1	5-3-1	1.082	13.691
				2	5-1	16.489	13.691
				3	5-2-1	3.131	13.690
	6-1	11.747	26.216	1	6-3-1	11.747	26.216
				2	6-3-5-1	0.0	30.929
	9-4	9.279	20.501	1	9-7-4	9.279	20.501
	9-5	20.702	13.691	1	9-5	12.940	16.320
				2	9-7-5	3.931	16.320
				3	9-8-5	3.832	16.320
	9-6	11.747	22.641	1	9-8-6	11.747	22.641
				2	9-5-8-6	0.0	29.080

$ERROR_1$ = 0%
$ERROR_2$ = 0.00%
$ERROR_3$ = 0.44%
$ERROR_4$ = 0.42%

*on an IBM 4032 using Fortran H, optimization level 3.

TABLE 5.10

RESULTS OF APPLYING ALG4, THE HYBRID ALGORITHM, TO EXAMPLE 1B,

RATE (comm. 1) = 1.8 * MARGINAL COST

RATE (comm. 2) = 1.6 * MARGINAL COST

$\varepsilon_1 = \varepsilon_2 = \varepsilon_3 = \varepsilon_4 = 5\%$

$\varepsilon_5 = \varepsilon_6 = \varepsilon_7 = \varepsilon_8 = 20\%$

$\varepsilon_9 = \varepsilon_{10} = \varepsilon_{11} = \varepsilon_{12} = 15\%$

Results of ALG2

No. of iterations for ALG2	=	5
No. of iterations for ALG3	=	32
CPU time, sec*	=	3.75
No. of shortest path calls	=	672

Results of ALG1

No. of iterations	=	5
CPU time, sec*	=	2.25
No. of shortest path calls	=	92

Totals

Total CPU time, sec*	=	6.00
Total no. of shortest path calls	=	764

$ERROR_1 = 0\%$ $ERROR_2 = 0.00\%$ $ERROR_3 = 0.67\%$ $ERROR_4 = 0.41\%$

*on an IBM 4031 using Fortran H, optimization level 3.

TABLE 5.11

RESULTS OF APPLYING ALG4, THE HYBRID ALGORITHM, TO EXAMPLE 1B

RATE (comm. 1) = 8.0 + 0.9 * MARGINAL COST

RATE (comm. 2) = 5.0 + 0.6 * MARGINAL COST

$\varepsilon_1 = \varepsilon_2 = \varepsilon_3 = \varepsilon_4 = 5\%$

$\varepsilon_5 = \varepsilon_6 = \varepsilon_7 = \varepsilon_8 = 20\%$

$\varepsilon_9 = \varepsilon_{10} = \varepsilon_{11} = \varepsilon_{12} = 15\%$

Results of ALG2

No. of iterations for ALG2	=	5
No. of iterations for ALG3	=	32
CPU time, sec*	=	3.74
No. of shortest path calls	=	672

Results of ALG1

No. of iterations	=	10
CPU time, sec*	=	5.18
No. of shortest path calls	=	236

Totals

Total CPU time, sec*	=	8.92
Total no. of shortest path calls	=	908

$ERROR_1 = 0\%$ $ERROR_2 = 0.00\%$ $ERROR_3 = 0.00\%$ $ERROR_4 = 0.01\%$

*on an IBM 4031 using Fortran H, optimization level 3.

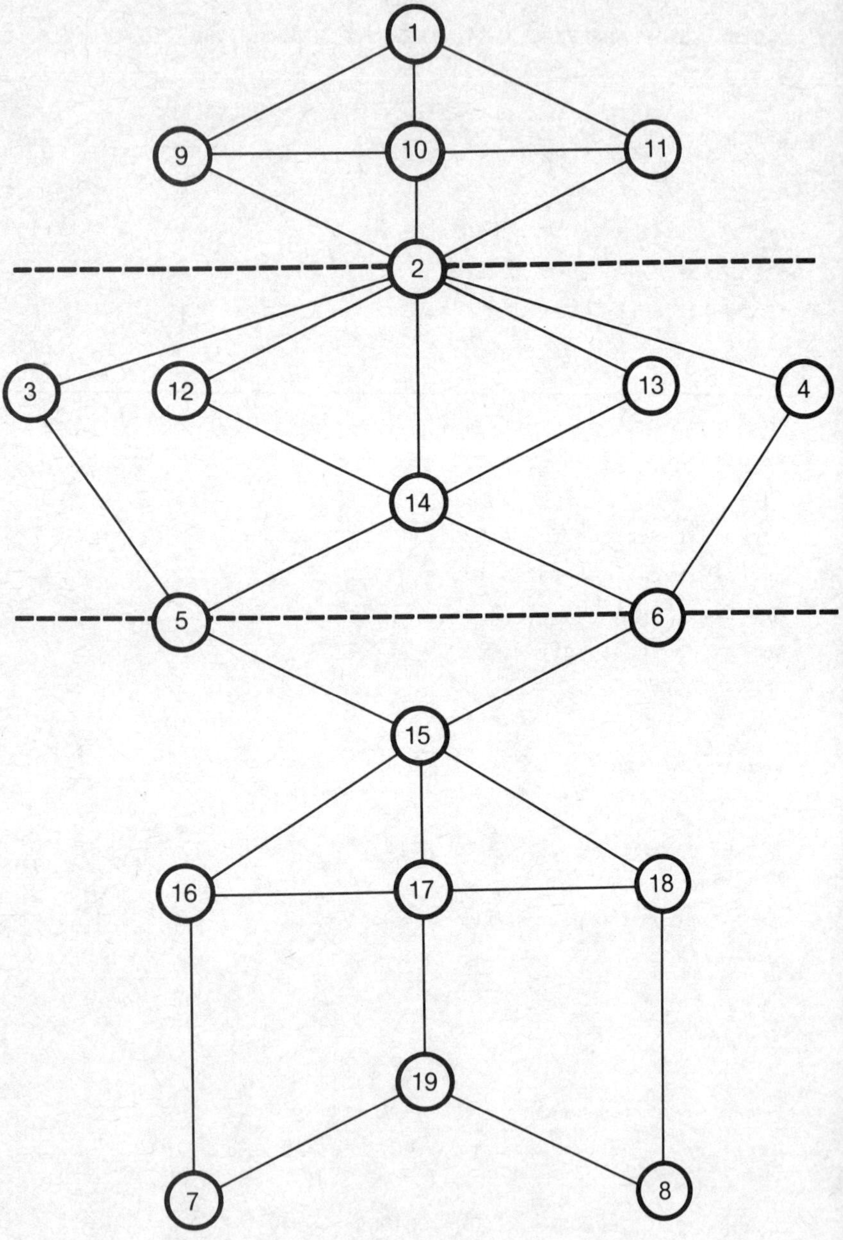

Figure 5.2: Carrier Network for Example 2

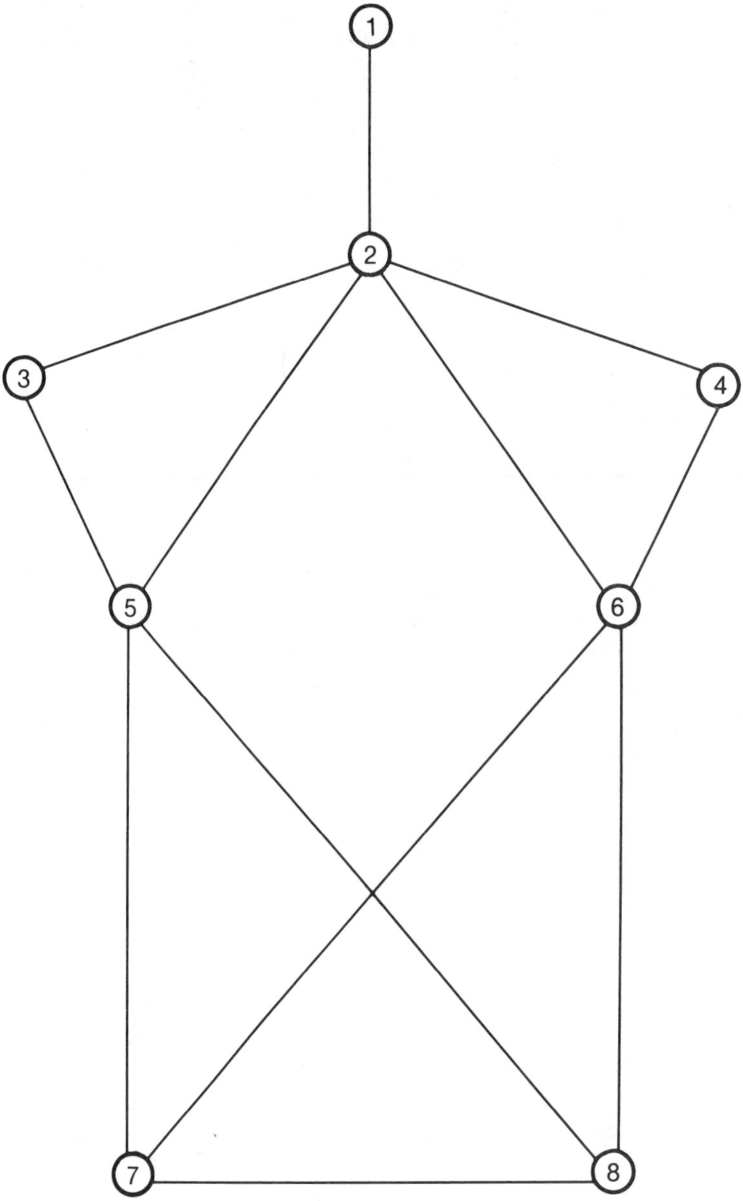

Figure 5.3: Shipper Network for Example 2

TABLE 5.12

DATE FOR EXAMPLE 2

No. of carriers	=	3
No. of carrier arcs	=	66
No. of carrier nodes	=	19
No. of shipper arcs	=	24
No. of shipper nodes	=	8

(other data available from the author by request)

TABLE 5.13

RESULTS OF APPLYING ALG2, THE DIAGONALIZATION/RELAXATION ALGORITHM, TO EXAMPLE 2

$\varepsilon_5 = \varepsilon_6 = \varepsilon_7 = \varepsilon_8 = 10\%$

$\varepsilon_9 = \varepsilon_{10} = \varepsilon_{11} = \varepsilon_{12} = 15\%$

No. of iterations for ALG2	=	2
No. of iterations for ALG3	=	17
CPU time, sec*	=	4.11
No. of shortest path calls	=	289

$ERROR_1$	=	94.38%
$ERROR_2$	=	6.12%
$ERROR_3$	=	35.18%
$ERROR_4$	=	2.06%

*on an IBM 4031 using Fortran H, optimization level 3.

TABLE 5.14

RESULTS OF APPLYING ALG4, THE HYBRID ALGORITHM, TO EXAMPLE 2

$\varepsilon_1 = \varepsilon_2 = \varepsilon_3 = \varepsilon_4 = 5\%$

$\varepsilon_5 = \varepsilon_6 = \varepsilon_7 = \varepsilon_8 = 20\%$

$\varepsilon_9 = \varepsilon_{10} = \varepsilon_{11} = \varepsilon_{12} = 10\%$

Results of ALG2

No. of iterations for ALG2	= 2
No. of iterations for ALG3	= 8
CPU time, sec*	= 2.22
No. of shortest path calls	= 136

Results of ALG1

No. of iterations	= 6
CPU time, sec*	= 5.47
No. of shortest path calls	= 96

Totals

Total CPU time, sec*	= 7.69
Total no.of shortest path calls	= 232

$ERROR_1 = 0.00\%$

$ERROR_2 = 0.00\%$

$ERROR_3 = 0.02\%$

$ERROR_4 = 0.03\%$

*on an IBM 4031 using Fortran H, optimization level 3.

From the above examples, it is quite evident that, at least for problems of the size presented, the complementarity-based approach is more efficient than the Frank-Wolfe-based algorithm. Theoretically, this result is expected due to the convergence rates of the two algorithms. However, there are two problems with ALG1. First, by having to store all active path definitions and flows, this approach is infeasible for very large problems. Secondly, since ALG3 exploits the network structure of the problem when solving the LP, and ALG1 does not when solving the linear complementarity subproblem, it is not clear that the CPU time efficiency of ALG1 will hold as the problems become larger. Even though ALG2 may take many more iterations to yield results of comparable accuracy, each iteration of ALG2 may possibly be done much more efficiently than an iteration of ALG1.

Let us now turn to consider alternative algorithms for variational inequalities and how they relate to GSPEM.

5.1 Alternative Algorithms

The algorithm which was presented in the previous section for the variational inequality/equivalent optimization formulation of GSPEM has as its main advantage a relatively small computer storage requirement. Given the very large memory requirements of the simple Frank-Wolfe algorithm (6 megabytes for the application discussed in Chapter 6), elaborate variational inequality/nonlinear programming algorithms which employ second-order information simply cannot be used. However, the sublinear convergence rate of the Frank-Wolfe algorithm is a well-known phenomenon. Thus, there exists a rather stark tradeoff between computational efficiency and the size of the problem which can be addressed.

The purpose of this section is to present two relatively simple extensions/variants of the diagonalization/Frank-Wolfe algorithm which, although not yet tested on GSPEM, should prove to be useful in the sense of increasing computational efficiency without dramatically increasing computer storage.

Consider the equivalent optimization problem (EOP) formulation of GSPEM (5.15) which was solved in the previous section by the Frank-Wolfe algorithm. The main problem which arises in the solution of (5.15) is the fact that there exists an enormous number of path variables h_p and g_q which simply cannot be enumerated in full. The main advantage of the Frank-Wolfe algorithm is that only one path per O-D pair is generated each iteration and these paths do not need to be stored from iteration to iteration. However, consider the LP subproblem (5.30) which results from the application of the Frank-Wolfe algorithm to problem (5.15). Given the structure of the constraints, the optimal solution to (5.30) will have many variables equal to their upper or lower bounds. To see this, simply note that if the upper and lower bound constraints in (5.30) were removed, this problem would have an unbounded solution. Thus, the solution to (5.30) will typically be at an extreme vertex. The final solution to (5.15) will typically not be near such a vertex and, thus, the Frank-Wolfe algorithm exhibits a 'bang-bang' behavior in which the LP subproblem finds extreme vertices and the linesearch procedure must then work hard to 'pull' the solution back into the center of the polyhedron. If the bounds placed on the LP subproblem were good in the sense of being near the optimal solution to (5.15), then this behavior of the Frank-Wolfe algorithm would not be troublesome. However, these bounds are typically ad hoc and thus very far from the optimum, leading to very poor performance of the algorithm.

The 'bang-bang' behavior of the Frank-Wolfe algorithm is well known in the solution of the standard traffic assignment problem; e.g., LeBlanc and Farhangian (1981). Evans (1976) devised a simple variant of this algorithm which overcame the 'bang-bang' problem by linearizing only a portion of the elastic demand traffic assignment problem. In what follows, we shall employ Evans's idea to solve GSPEM.

Consider the nonlinear program (5.15). Instead of linearizing the entire objective function Z, as in (5.24), let us select the feasible direction y^i in iteration i by linearizing only the function in the last two summations in (5.15) to yield the following subproblem:

$$\text{minimize} \quad \tilde{Z} = \sum_{\ell \in L} \left[\int_0^{S_\ell} \Psi_\ell(s,S^k)d_s - \int_0^{D_\ell} \Theta_\ell(s,D^k)d_s \right]$$

$$\sum_{a \in A} \Phi\, t_a^i\, f_a + \sum_{b \in B} MC_b^i\, e_b \qquad (5.38)$$

$$\text{subject to:} \quad x = (s;\ D;\ f;\ e) \ \varepsilon\ \Omega^i\ .$$

As before, (5.38) will have as part of its optimal solution the characteristic of an all-or-nothing assignment of flows on the shortest paths between each shipper and carrier O-D pair. Thus, (5.38) can be rewritten using (5.26) and (5.29) as:

$$\text{minimize} \quad \tilde{Z} = \sum_{\ell \in L} \left[\int_0^{S_\ell} \Psi_\ell(s,S^k)d_s - \int_0^{D_\ell} \Theta_\ell(s,D^k)d_s \right]$$

$$\sum_{w \in W} TC_w^* \, T_w \qquad (5.39)$$

subject to:

$$\sum_{\ell \in L} D_\ell - \sum_{\ell \in L} S_\ell = 0$$

$$S_\ell - D_\ell + \sum_{i \in L} \sum_{(i,\ell) \varepsilon W} T_{(i,\ell)} - \sum_{j \in L} \sum_{(\ell,j) \varepsilon W} T_{(\ell,j)} = 0 \quad \forall\ \ell \varepsilon L,$$

$$x^i - U_1 \leq x \leq x^i + U_2$$

$$x \geq 0.$$

Problem (5.39) constitutes a nonlinear network flow model and, although not as easy to solve as (5.24), many efficient algorithms exist for the solution of this problem; e.g., Dembo (1983) and Klincewicz (1983). Thus, each subproblem (5.39) will take longer to solve than the linear subproblem (5.24), but, as the experience of LeBlanc and Farhangian (1981) shows, far fewer iterations should be necessary. To see this, one need only note that if the marginal costs MC_b and arc delays $\tilde{t}_a$ were constants, then the solution of (5.39) would yield an exact solution to the original problem (5.15). In practice, these functions are not equal to constants, but there are significant portions of the function which are constant; e.g., Figures 6.5 and 6.6 in the next chapter. Very few arcs are normally congested in real freight transportation networks, and thus the solution to (5.39) should significantly improve the performance of the feasible direction algorithm for GSPEM.

Harker (1985a) presents the details of implementing this variant of the Frank-Wolfe algorithm on the pure spatial price equilibrium model; i.e., problem (5.15) without the last summation and the carrier path and arc flow constraints. In addition to the theoretical development of this algorithm, a series of test problems was run in order to compare this new algorithm with the Frank-Wolfe algorithm. As Table 5.15, illustrates, very significant savings in computational speed are found with this Evans-type algorithm. Although significantly faster, the new algorithm does not require any significant increase in computer storage over the Frank-Wolfe algorithm. Also, the only modification to the previous software is to replace the LP code to solve (5.24) with a nonlinear network code to solve (5.39).

Therefore, the simple variant of the Frank-Wolfe feasible direction algorithm to solve (5.15) has proven to be very easy to implement and computationally superior on the standard spatial price equilibrium problem, and should also outperform Frank-Wolfe in GSPEM.

The second variant of the algorithm presented in the previous section deals with recent developments in what is known as simplicial decomposition algorithms for variational inequalities. The diagonalization/relaxation algorithm for variational inequalities relies on the ability to solve the mathematical programming subproblems efficiently. The simplicial decomposition approach does not employ the same strategy, but relies on the fact that the solution to a variational inequality is usually near to one or a few of the vertices of the polyhedron which constitutes the feasible region. Using this property, the iterations of the simplicial decomposition algorithm entail the generation of vertices and use these vertices to form an approximate solution to the variational inequality. Thus, the algorithm is a vertex generation procedure rather than an iterative method like diagonalization/relaxation.

Consider the following variational inequality:

$$\text{find } x^* \in X$$

where

$$F(x^*)^T \; (x - x^*) \geq 0 \qquad (5.40)$$

where $F{:}R^n \rightarrow R^n$ and $X \subseteq R^n$ is a compact polyhedral set. Let x^i be the ith vertex of X and let α be the index of the vertices $\alpha = \{i \mid x^i \text{ is a vertex of } X\}$. Using barycentric coordinates, any point $x \in X$ can be written as:

$$x = \sum_{i \in \alpha} x^i \pi_i \qquad (5.41)$$

where π_i are scalars which obey

$$\sum_{i \in \alpha} \pi_i = 1 \qquad (5.42a)$$

$$\pi_i \geq 0 \qquad \forall \; i \in \alpha \, . \qquad (5.42b)$$

The essential idea behind the simplicial decomposition algorithm is to use (5.40) to generate a new extreme points and then solve a lower dimensional variational inequality problem in terms of the π_i variables. Let us detail this procedure.

TABLE 5.15

COMPARISON OF THE EVANS-TYPE AND FRANK-WOLFE ALGORITHMS FOR PURE SPATIAL PRICE EQUILIBRIUM*

	CPU Time (sec) on DEC 10/90 to Achieve the Same Minimum Value of the Objective Function Z		% of Savings in CPU Time
Example	Frank-Wolfe	Evans-type	
1	12.55	1.98	84.2
2	60.70	34.49	43.2
3	35.18	7.43	79.0
4	193.06	8.53	95.6

*from Table 6 in Harker (1985a).

Let α^k be the index set of vertices which have been generated up to and including iteration k and let x^k be the current approximate solution to (5.40)

$$x^k = \sum_{i \varepsilon \alpha^k} \pi_i^k x^i \qquad (5.43)$$

where π_i^k are the barycentric weights at iteration k. Consider (5.40) with $x^* = x^k$:

$$F(x^k)^T x \geq F(x^k) x^k . \qquad (5.44)$$

If (5.44) holds for all $x \varepsilon X$, then x^k is the solution to (5.40). If not, then there must exist at least one $x \varepsilon X$ such that the inequality in (5.44) is reversed. Such a point can be found by solving the following linear program:

$$\underset{x \varepsilon X}{\text{minimize}} \; F(x^k) \; x. \qquad (5.45)$$

Obviously, the solution to (5.45) will generate a new vertex of X. Using this new vertex, one can augment the set α^k to form α^{k+1} and compute x^{k+1} by solving:

$$F(x^{k+1}) \; (x - x^{k+1}) \geq 0 \qquad \forall \quad x \varepsilon X^{k+1} \qquad (5.46)$$

where x^{k+1} is the convex hull of the α^{k+1} vertices

$$X^{k+1} = \{x: \; x = \sum_{i \varepsilon \alpha^{k+1}} x^i \pi_i$$

$$\sum_{i \varepsilon \alpha^k} \pi_i = 1$$

$$\pi_i \geq 0 \qquad \forall \quad i \; \varepsilon \; \alpha^k\} .$$

The dimension of this problem (5.46) is clearly the number of vertices which have been generated $|\alpha^{k+1}|$. Therefore, (5.46) should be much easier to solve than the original variational inequality (5.40).

In summary, the basic steps of the simplicial decomposition algorithm are:

Step 0 Generate an initial vertex of X, x^1, by solving (5.45) and let $k = 1$.

Step 1 Find $x^k \in X^k$ such that

$$F(x^k)^T(x-x^k) \geq -\varepsilon_k \quad \forall \quad x \in X^k \tag{5.47}$$

Step 2 Solve (5.45) for x^{k+1}. If the gap

$$G(x^k) = F(x^k)(x^{k+1}-x^k) = \pm \varepsilon_k , \tag{5.48}$$

stop; x^k is a solution to (5.40). Else, let $\alpha^{k+1} = \alpha^k \cup (k{+}1^{st}$ vertex $x^{k+1})$, increment k and return to Step 1.

At each iteration a new vertex is generated and must be stored. It is possible to drop some previously generated vertices, as discussed by Lawphonpanich and Hearn (1984), in order to keep the dimensionality of (5.47) low, but several vectors of dimension n must be stored. This fact points out the main weakness of simplicial decomposition in the present context. Both Lawphongpanich and Hearn (1984) and Pang and Yu (1984) have shown that in the context of the traffic assignment problem, the simplicial decomposition algorithm is computationally far superior to the standard Frank-Wolfe approach. However, it is unclear what the memory requirements of this algorithm will be on realistically sized problems. As was stated previously, the implementation of the Frank-Wolfe algorithm, in which only two vectors of dimension n need to be stored, took 6 megabytes of memory for the application in Chapter 6. Whether or not this simplicial decomposition approach can be used effectively in solving GSPEM remains an open research question.

In order to implement this algorithm for GSPEM, one must first be sure to place upper bounds on all supplies and demands in order to make the feasible region Ω in (5.9) a compact polyhedral set. Given these bounds, the solution of (5.45) is identical to the solution of (5.24), which can be accomplished with any network LP algorithm. The only complicated step in the simplicial decomposition implementation is in solving (5.47). Any variational inequality algorithm can be employed, although the linearization technique discussed in Pang and Chan (1982) and used in Pang and Yu (1984) in the context of the traffic assignment problem seems the most promising technique, since it generates very low dimension quadratic programs which can be solved very efficiently with the Dantzig-van de Panne-Whinston algorithm (Pang, 1981c).

Therefore, two potentially useful extensions of the algorithm presented for GSPEM in the previous section have been discussed and, if they can be implemented without greatly increasing the computer storage requirements, they should both prove to be superior to the diagonalization/Frank-Wolfe approach. Future research on the efficient solution of GSPEM along these lines is vital if this model is to be used effectively.

Chapter 6
APPLICATION OF GSPEM TO THE U.S. COAL ECONOMY

In this chapter, a large-scale application of the Generalized Spatial Price Equilibrium Model (GSPEM) to the U. S. coal economy is presented. The purpose of this application is two-fold First, this application shows that GSPEM is a viable tool for large-scale applications. That is, GSPEM is not restricted to problems of small to moderate size, but can be used on realistically large problems. Second, this application shows how GSPEM can be used as a tool to address certain policy issues.

The U. S. coal economy has received much attention in recent years. The models by Baughman, Joskow and Kamati (1979), Zimmerman (1981) and Shapiro and White (1982) contain detailed analyses of the supply and demand for coal, and also contain a simple transportation submodel to predict interregional flows. In the work by Friesz, et al. (1981), simple (fixed) supplies and demands of coal are used with a more sophisticated transportation model. Osleeb and Ratick (1982) have developed a model of coal movement which focuses on the question of how port expansions will affect the regional movements of coal to export points.

In GSPEM, the effect of the decisions of producers, consumers, shippers and transportation firms on the price, supply, demand and regional distribution of coal can be taken into account. In this chapter, GSPEM will be run assuming marginal cost pricing so that the optimization-based algorithm (ALG3) can be used. The model will be run with 1980 data and the results (the base-year predictions) will be compared against the historical data for 1980. Then, a scenario in which exports through certain ports are doubled in volume and other ports are closed is addressed in order to show the potential usefulness of GSPEM as a policy analysis tool.

The next section describes the assumptions which underlie the use of the optimization-based solution algorithm used on this application. Section 6.3 describes the regional descriptions which are used and the estimation of the regional supply and demand functions. The carrier network which was used is described in Section 6.4, and the description of the shipper network is presented in Section 6.5. The base-year predictions are presented in Section 6.6, Section 6.7 presents the export scenario results, and conclusions are drawn in the last section of this chapter.

6.1 Assumptions Underlying the Optimization-Based Algorithm

As was discussed in Chapter 5, GSPEM can be solved by either a complementarity-based algorithm (ALG1) or an optimization-based algorithm (ALG2). ALG1 has the problem of requiring a substantial amount of computer storage, and thus is currently infeasible for problems of the magnitude addressed in this application.

The optimization-based algorithm, ALG2, however, needs two assumptions to be applicable. First, marginal cost pricing ($R_v = MC_v^* \; \forall \; v \in V$) must be assumed for all carrier origin-destination (O-D) pairs. Second, the time delays cannot be generated from flows on the carrier network, but must be stated as a function of the flows on the shipper network (exogenous aggregation).

Given these two assumptions, ALG2 was implemented using the out-of-kilter algorithm to solve the network linear programming subproblem and using Powell's (1983) implementation of Dial, Glover, Karney and Klingman's (1979) label-correcting shortest path algorithm C2. Finally, a bisection linesearch was used, but since some functions are nonmonotonic, this linesearch procedure has been modified to insure that a local minimum is reached at each iteration. Basically, the modification is one in which the neighborhood around the solution to the linesearch procedure is checked to see that the objective function is convex in this neighborhood.

All of the work presented in this chapter was performed on Argonne National Laboratory's IBM 3033 using the Fortran H compiler.

6.2 Data Description

The regional definitions which will be considered in the application presented in this chapter are the 181 Bureau of Economic Analysis (BEA) regions comprising the 48 contiguous states. These BEA regions are shown in Figure 6.1. Thus, there will be, in general, 181 supply and demand regional considered in this application.

The data to estimate the regional supply and demand functions are difficult to obtain, so therefore, we have resorted to a 'second-best' approach. Basically, national supply and demand functions for coal (total volume, not disaggregated by BTU content and sulfur-content) have been estimated, and then disaggregated by 1980 regional supply and demand data from Reebie Associates (1981). Let us describe in greater detail this estimation/disaggregation procedure.

To begin, a national supply function for coal and a national demand function for coal were estimated. The functions which were estimated took the form:

$$D^t = a + b\ DP^t_{coal} + c\ P^t_{oil} + d\ P^t_{gas} + e\ GNP^t \quad (6.1)$$

$$S^t = \alpha + \beta\ P^t_{coal} + \gamma\ PROD^t + \lambda\ STOCK^{t-1}, \quad (6.2)$$

where

D^t = the U. S. demand for coal at time t,

DP^t_{coal} = the delivered price of coal at the powerplants at time t (a proxy for the total demand price of coal since 81.93% of the total consumption of coal in the U. S. is by powerplants (Table 55 of Energy Information Administration, 1982),

P^t_{oil} = the price of crude oil at time t,

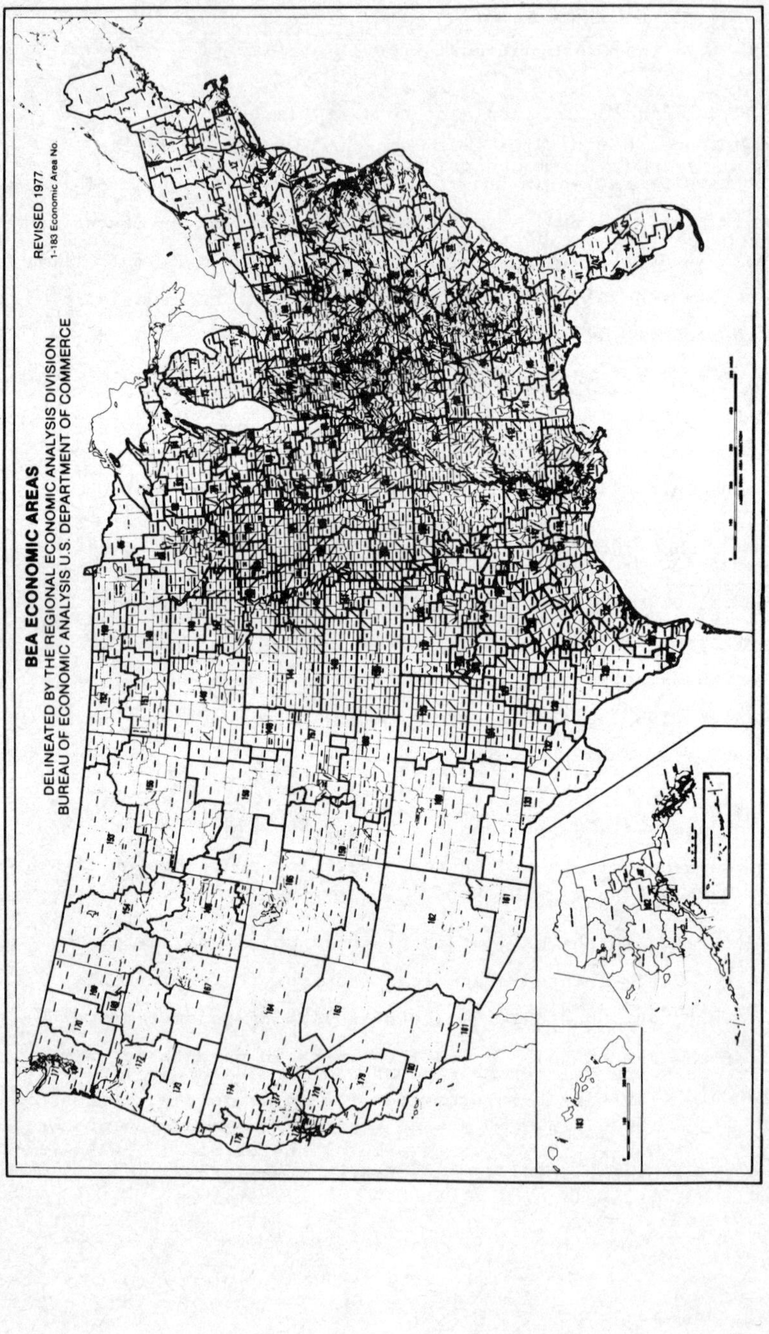

Figure 6.1: BEA Economic Areas

P^t_{gas} = the price of natural gas at time t,

GNP^t = the gross national product at time t,

S^t = the U. S. supply of coal at time t,

P^t_{coal} = the mine-mouth price of coal at time t,

$PROD^t$ = labor productivity, average short-tons per man-day, at time t,

$STOCK^t$ = the total stock of coal at time t.

Letting

E^t = the total U. S. net exports at time t,

$CTRANS^t$ = the cost of transport at time t,

we can see that (6.1) - (6.2) constitute a simultaneous equation system, where

$E^t = S^t - D^t$ and

$CTRANS^t = DP^t_{coal} - P^t_{coal}$.

That is, by assuming $S^t = D^t + E^t$ at each time period t, (6.1) -(6.2) endogenously determines P^t_{coal}, and thus must be estimated simultaneously (see Johnston, 1972, for a discussion of simultaneous equation systems).

The data used in the estimation of (6.1)-(6.2) are listed in Table 6.1. Using a three-stage least squares regression (see Kmenta, 1971, or Johnston, 1972), the following parameters were found for the system (6.1) - (6.2) (numbers in parentheses are the standard errors):

$$D^t = \underset{(41.169)}{217.308} - \underset{(2.441)}{3.460DP^t_{coal}} + \underset{(0.680)}{0.747P^t_{oil}} + \underset{(1.946)}{3.675P^t_{gas}} + \underset{(0.024)}{0.276GNP^t} \qquad (6.3)$$

TABLE 6.1

DATA FOR ESTIMATION OF U.S. SUPPLY AND DEMAND FUNCTIONS FOR COAL

Year^t	S^{t1}	D^{t2}	P_{coal}^{t3}	DP_{coal}^{t4}	P_{oil}^{t5}	P_{gas}^{t6}	GNP^{t7}	$PROD^{t8}$	$STOCK^{t9}$
1961	420.4	390.3	9.53	12.89	16.11	4.46	756.6	13.87	73.0
1962	439.0	402.2	8.99	12.08	15.87	4.51	800.12	14.72	71.3
1963	477.2	433.5	8.55	11.41	15.58	4.53	832.6	15.83	71.5
1964	504.2	445.7	8.41	10.84	15.29	4.37	876.4	16.84	76.7
1965	527.4	427.3	8.03	10.33	14.86	4.34	929.3	17.52	78.6
1966	546.8	497.7	7.70	9.77	15.40	4.20	984.92	18.52	75.6
1967	564.9	491.4	7.37	9.36	14.27	4.17	1011.4	19.17	94.6
1968	556.7	509.8	6.86	8.70	13.76	4.05	1058.1	19.37	87.0
1969	571.0	516.4	6.63	8.13	13.76	3.95	1087.7	19.90	81.9
1970	612.7	523.2	7.49	8.53	13.44	3.85	1085.6	18.84	93.0
1971	560.9	501.6	7.67	8.68	13.64	3.91	1122.4	18.02	91.0
1972	602.5	524.3	7.66	8.44	13.10	3.84	1185.9	17.74	116.8
1973	598.6	562.6	7.64	8.06	14.22	4.26	1255.0	17.58	104.6
1974	610.0	558.4	11.93	11.70	23.10	5.48	1248.02	17.58	96.6
1975	654.6	562.6	12.20	11.18	23.60	7.35	1233.8	14.74	128.3
1976	684.9	603.8	11.13	10.53	23.95	9.13	1300.5	14.46	134.7
1977	697.2	625.2	10.13	10.42	23.68	11.75	1371.7	14.84	157.3
1978	670.2	625.2	9.68	10.55	23.17	12.60	1436.9	14.68	145.9
1979	781.1	680.5	8.93	9.87	30.00	15.07	1483.02	15.23	182.0
1980	829.7	702.7	7.81	9.15	56.16	18.70	1480.7	15.88	204.0
1981	807.7	727.7	7.00	8.19	62.22	22.14	1518.2	16.12	182.5

1. From Energy Information Administration (1982), Table 53, million short tons.
2. Ibid., Table 53, million short tons.
3. Ibid., Table 63, dollars per short tons, 1972 dollars.
4. Ibid., Table 55, dollars per short tons, 1972 dollars.
5. Ibid., Table 39, dollars per equivalent short tons, 1972 dollars, where conversions to equivalent short tons are done by the conversion factors on p. 223. of EIA (1982).
6. Ibid., Table 51, dollars per equivalent short tons, 1972 dollars.
7. From U. S. Statistical Abstracts.
8. EIA (1982), Table 59, average short tons per man-day.
9. Ibid., Table 57, million short tons.

$$S^t = \underset{(216.553)}{90.765} + \underset{(8.368)}{10.594P^t_{coal}} + \underset{(8.402)}{8.778PROD^t} + \underset{(0.305)}{2.633STOCK^{t-1}} \tag{6.4}$$

weighted R^2 for system = 0.973.

Using the 1980 values for all variables but D, S, and prices, the supply and demand functions are

$$D^{1980} = 728.435 - 3.460DP^{1980}_{coal} \tag{6.5}$$

$$S^{1980} = 709.404 + 10.594P^{1980}_{coal} . \tag{6.6}$$

Figure 6.2 shows the above functions, plus the function $(D^{1980} + E^{1980}) = D^{1980} + 127.0$.

To disaggregate the above functions, data from Reebie Associates (1981) on 1980 supplies, S^{1980}_{ℓ}, and demands, D^{1980}_{ℓ}, for all BEA regions $\ell \in L$ (the total set of regions are) used. From this data, the regional supply and demand functions are derived as follows:

$$S^{1980}_{\ell} = (\bar{S}^{1980}_{\ell}/\bar{S}^{1980}) \cdot (709.404 + 10.594\ P^{1980}_{coal}) \tag{6.7}$$

$$D^{1980}_{\ell} = (\bar{D}^{1980}_{\ell}/\bar{D}^{1980}) \cdot (728.435 - 3.460\ DP^{1980}_{coal}) , \tag{6.8}$$

where $\bar{S}^{1980}$ and $\bar{D}^{1980}$ are the total supply and demand in the nation in 1980.

The model just presented for regional supply and demand functions is extremely simple, and the estimations of the national functions have high standard errors. Therefore, these functions should not be used in actual policy analysis. However, the purpose of the applications presented in this chapter is not to perform detailed policy analysis, but rather to show how GSPEM can handle realistic problems and how it can be used for policy analysis. Therefore, the functions derived via (6.7) - (6.8) are

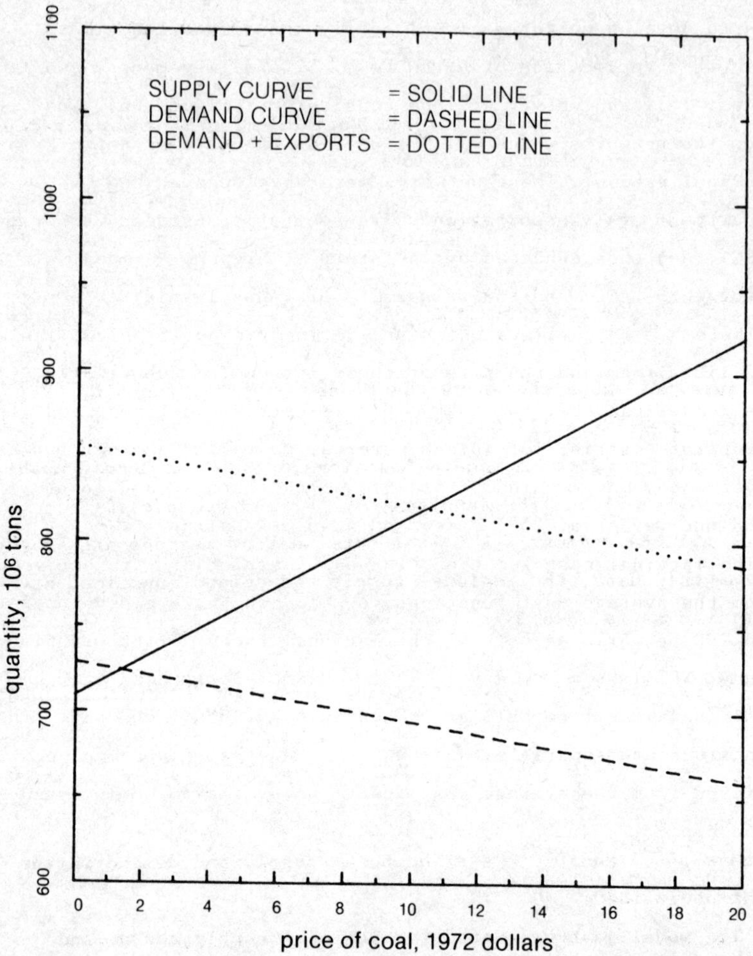

Figure 6.2: Estimated U.S. Supply and Demand Curves for Coal

not to be considered real, but rather 'realistic.' Better data and more detailed analyses of coal supply and demand would allow one to develop better regional supply and demand functions.

In this application of GSPEM to U. S. coal movements, two modes, rail and water, will be considered. Figures 6.3 and 6.4 show the networks which will be used as the carrier network in our application. These networks were developed at the U. S. Department of Transportation's Transportation Systems Center (TSC), and then enhanced by the staff at Argonne National Laboratory (ANL) and the University of Pennsylvania.

There are 18 carriers which are defined in our problem; Table 6.2 lists them and the size of their respective subnetworks. As the reader can see, there are 16 rail carriers defined, and also a separate carrier for inland waterway moves and another for deep draft water moves. The total size of the carrier network is 7688 one-way arcs, 2577 nodes and 4245 O-D pairs.

The marginal cost functions for each carrier arc are derived from the average cost functions, $AC_b(e)$, which were developed for the TSC networks as part of the National Energy Transportation Study (NETS) (see Bronzini, 1979). Examples of these average cost functions from NETS are shown in Figures 6.5 and 6.6. These functions are separable, and thus the marginal costs can be derived from the average cost functions by the following relationship:

$$MC_b(e) = AC_b(e_b) + e_b \, dAC_b/de_b \; . \qquad (6.9)$$

The shipper network was developed by first creating arcs to represent each carrier O-D pair. Arcs were then added to connect the 181 centroid nodes representing the BEA regions with the carriers' subnetworks. Finally, arcs were added to represent the transshipment between rail and water and interlinings between carriers. The final result is a network with 960 nodes, 6993 one-way arcs, and 1238 O-D pairs.

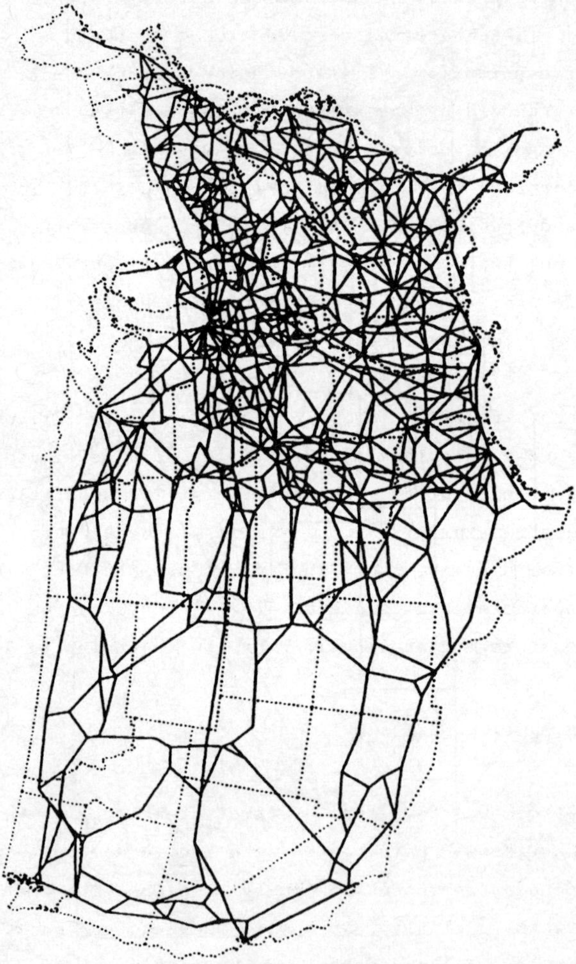

Figure 6.3: DOT Transportation Systems Center Rail Network

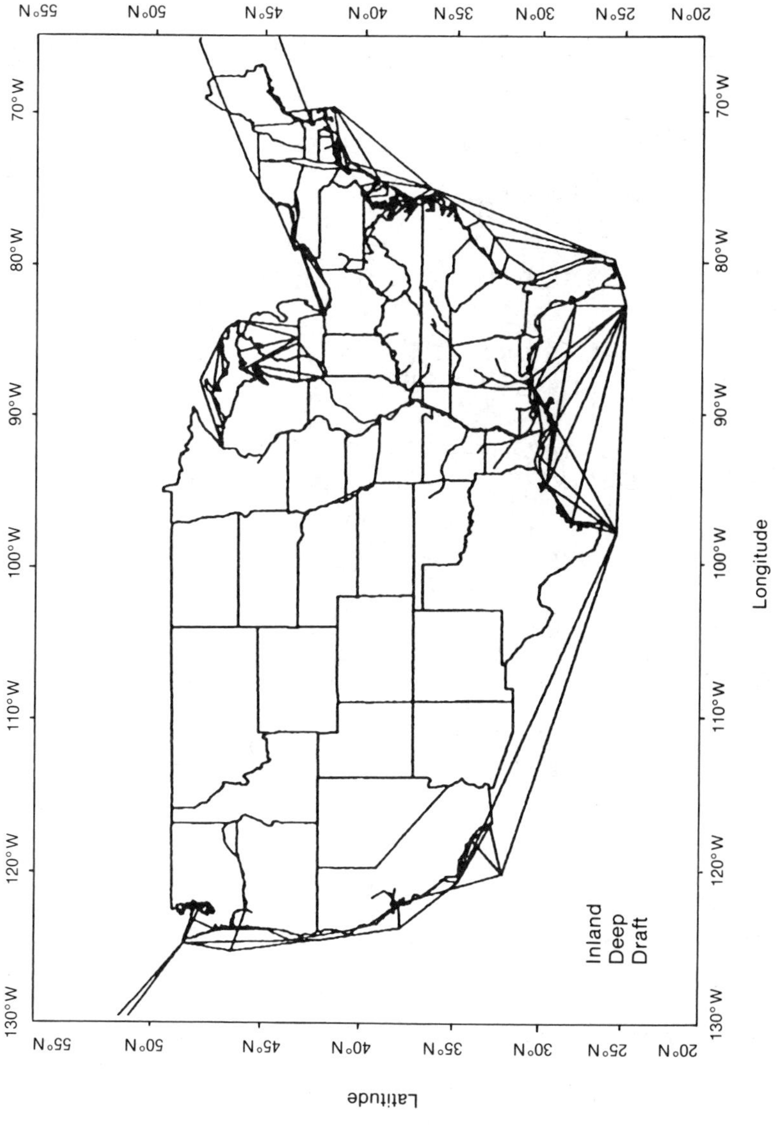

Figure 6.4: TSC Waterways Network

TABLE 6.2

CARRIER DESIGNATIONS AND NETWORK SIZES

Carrier No.	Carrier Name	No. Arcs	No. Nodes	O-D Pairs
1	Santa Fe R.R.	410	132	158
2	New England R.R.s	30	13	24
3	Burlington Northern R.R.	816	269	410
4	CSX Corporation	1160	334	530
5	Conrail	878	238	474
6	Denver and Rio Grande R.R.	16	8	12
7	Illinois Central Gulf R.R.	462	146	204
8	Kansas City Southern R.R.	82	32	20
9	Missouri, Kansas, Texas R.R.	142	57	42
10	Pacific Railways	654	220	276
11	Norfolk and Western and Southern R.R.	982	291	434
12	Southern Pacific R.R.	216	83	100
13	Chicago Northwestern R.R.	288	90	230
14	Milwaukee R.R.	290	104	160
15	Soo Line	32	19	8
16	Florida East Coast R.R.	10	5	4
17	*Inland Waterway	992	456	341
18	*Deep Draft Waterway	208	80	818
TOTAL		7668	2577	4245

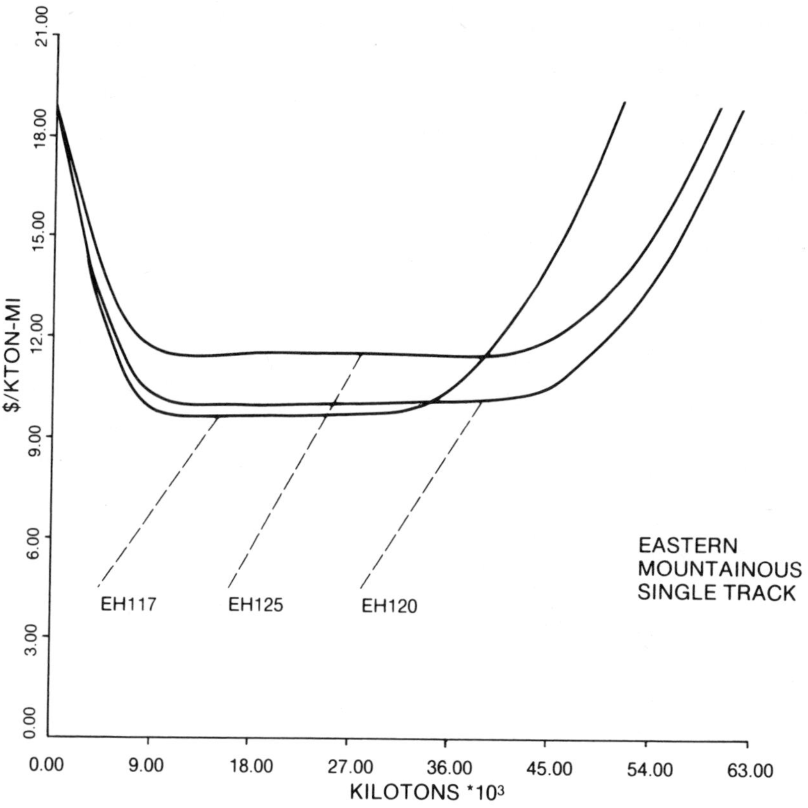

RAIL LINEHAUL LINK COST FUNCTIONS, EAST

*from CACI (1980)

Figure 6.5: Typical TSC Rail Average Cost Functions

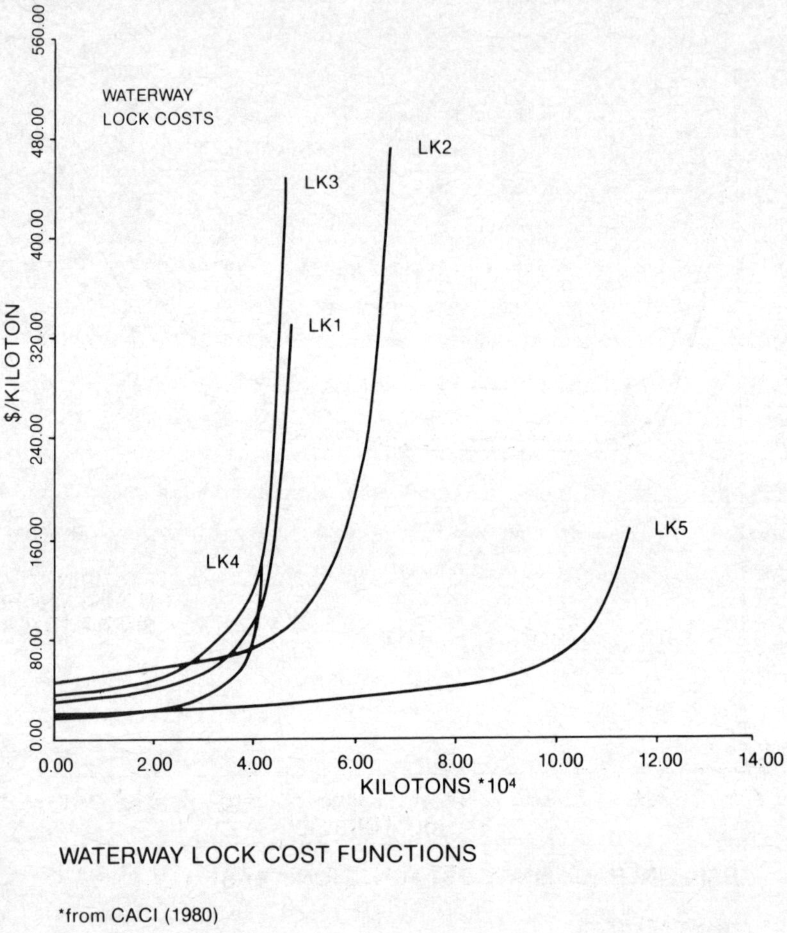

Figure 6.6: Typical TSC Water Average Cost Functions

Since one would expect a relatively low value of time for coal movements, a simplifying assumption that the value of time equals zero was made. Therefore, no time delay functions were employed on this shipper network since the value of time is zero.

Table 6.3 shows the tonnage of coal which was exported in 1980 from the various U. S. ports listed (from Maritime Administration, 1982). In the application presented in this chapter, exports are treated as fixed values.

Table 6.4 summarizes the database which was used in the base-year run of GSPEM and in the export scenario. As discussed in this table, the data from Reebie Associates covers only about 74% of the total U. S. coal supply and demand. The applications presented in this chapter will only consider this 74% due to the lack of O-D data for the remaining 26%. That is, the Reebie database will be used to compare predicted versus actual shipper O-D flows and regional supplies and demands; thus, we will only attempt to predict the 74% for which we have historical data.

Let us now begin our discussion of the base year predictions.

6.3 Base Year Predictions

In this section, the results of applying the optimization-based algorithm to the 1980 national coal database described in the previous sections are presented. Since the functions which are used in this application are separable, the diagonalization algorithm ALG2 presented in Chapter 5 collapses into a single application of ALG3, the Frank-Wolfe-based feasible direction algorithm. Again, marginal cost pricing ($R_v = MC_v^* \; \forall \; v \in V$) must be assumed.

Let us first make a few definitions. Absolute error is the maximum change in the variables from one iteration of ALG3 to the next, or, for iteration $i + 1$, absolute error can be stated as:

TABLE 6.3

1980 COAL EXPORTS*

Port	1980 Exports, million tons
New Orleans	3.825
Hampton Roads	51.711
Baltimore	12.126
Philadelphia[1]	2.966
New York	0.254
Toledo[2]	16.496
Mobile	2.444
TOTAL	89.822

*from Maritime Administration (1982)

[1]represents Delaware River category

[2]represents Great Lakes category

TABLE 6.4

SUMMARY OF THE DATABASE

Carrier Network

Number of Carriers	=	18
Number of Nodes	=	2577
Number of Arcs	=	7668
Number of O-D Pairs	=	4245

Shipper Network

Number of Nodes	=	960
Number of Arcs	=	6993
Number of O-D Pairs	=	1238

Total 1980 Supply of Coal	=	614.528 million tons[1]
Total 1980 Demand for Coal	=	524.706 million tons
Total 1980 Exports of Coal	=	89.822 million tons

[1] The supply and demand data are derived from the Reebie Associates (1981) database and do not contain the total U. S. tonnage, but approximately 74% of the total.

$$\text{absolute error} = \max\{E_1, E_2, E_3, E_4\}$$

where

$$E_1 = \max_{\ell \varepsilon L} \{|S_\ell^{i+1} - S_\ell^i|\}$$

$$E_2 = \max_{\ell \varepsilon L} \{|D_\ell^{i+1} - D_\ell^i|\}$$

$$E_3 = \max_{a \varepsilon A} \{|f_a^{i+1} - f_a^i|\}$$

$$E_4 = \max_{b \varepsilon B} \{|e_b^{i+1} - e_b^i|\} .$$

The stopping tolerance for ALG3 is

$$\text{absolute error} \leq 5 \times 10^4 \text{ tons, or}$$

$$\varepsilon_9 = \varepsilon_{10} = \varepsilon_{11} = \varepsilon_{12} = 5 \times 10^4 \text{ tons.}$$

Next, two measures of goodness of fit of the predicted, x^P, with the historical, x^H, are used. The first, normalized mean absolute error (MAE), is given by

$$\frac{1}{N} \sum_{i=1}^{N} \frac{|x_i^P - x_i^H|}{\bar{x}^P} \qquad (6.10$$

where

$$\bar{x}^P = \frac{1}{N} \sum_{i=1}^{N} x_i^P . \qquad (6.11)$$

Therefore, normalized MAE is a measure of how far the predicted is from the historical, with a value of zero being a perfect fit.

The second measure is the coefficient of determination, or R^2, which is given by

$$R^2 = 1 - \frac{\sum_{i=1}^{N} (x_i^H - x_i^P)^2}{\sum_{i=1}^{N} (x_i^H - \bar{x}^P)^2} . \quad (6.12)$$

Thus, a value of $R^2 = 1$ denotes a perfect fit.

The base year run is started by preloading the networks with all commodity flows except coal flows; this preloaded flow, called the background flow, is taken from the work by Gottfried (1983) to develop a national model of commodity flows. Thus, the implicit assumption is that all other commodity flows will remain constant as the coal flows equilibrate.

Table 6.5 summarizes the results of the base year run. The algorithm took 38 iterations to achieve a solution, and took 33 CPU minutes. Figure 6.7 shows the convergence of the algorithm in terms of the change in absolute error at each iteration; the 'zig-zagging' behavior depicted in Figure 6.7 is typical of Frank-Wolfe-based algorithms.

As Table 6.5 shows, the results of this run in terms of the fit of the supplies and demands is very good. In terms of shipper O-D flows, the total predicted volume moving throughout the network is fairly close to the actual volume (-9.32% difference), but the goodness of fit measures show poor results. Upon closer inspection, it appears that the result of poor O-D flow replication is a result of considering coal as a single commodity. In reality, there are many types of coal delineated by BTU and sulfur contents. The model predicts, in general, coal coming into a region from its neighboring region, whereas in reality the flows are much more dispersed due to the variations in BTU and sulfur contents. With better data, the various coal types could be defined and the predicted O-D flows would likely improve. Also, improvements to the commodity market model as outlined in the next chapter may also yield significantly better results. Simply stated, further empirical and theoretical work is needed.

TABLE 6.5

RESULTS OF BASE YEAR RUN

CPU time	= 33.05 minutes*
Number of iterations in ALG3	= 38

Supplies

1980 total supply	= 614.528 million tons
Predicted total supply	= 574.078 million tons
% difference	= -6.58%
Normalized MAE	= 0.101
R^2	= 0.989

Demands

1980 total supply	= 524.706 million tons
Predicted total supply	= 484.253 million tons
% difference	= 7.71%
Normalized MAE	= 0.085
R^2	= 0.990

Shipper O-D Flows

1980 total volume of flow	= 44.028 million tons
Predicted total volume of flow	= 39.925 million tons
% difference	= -9.32%
Normalized MAE	= 1.026
R^2	= 0.162

*on an IBM 3033, Fortran H compiler

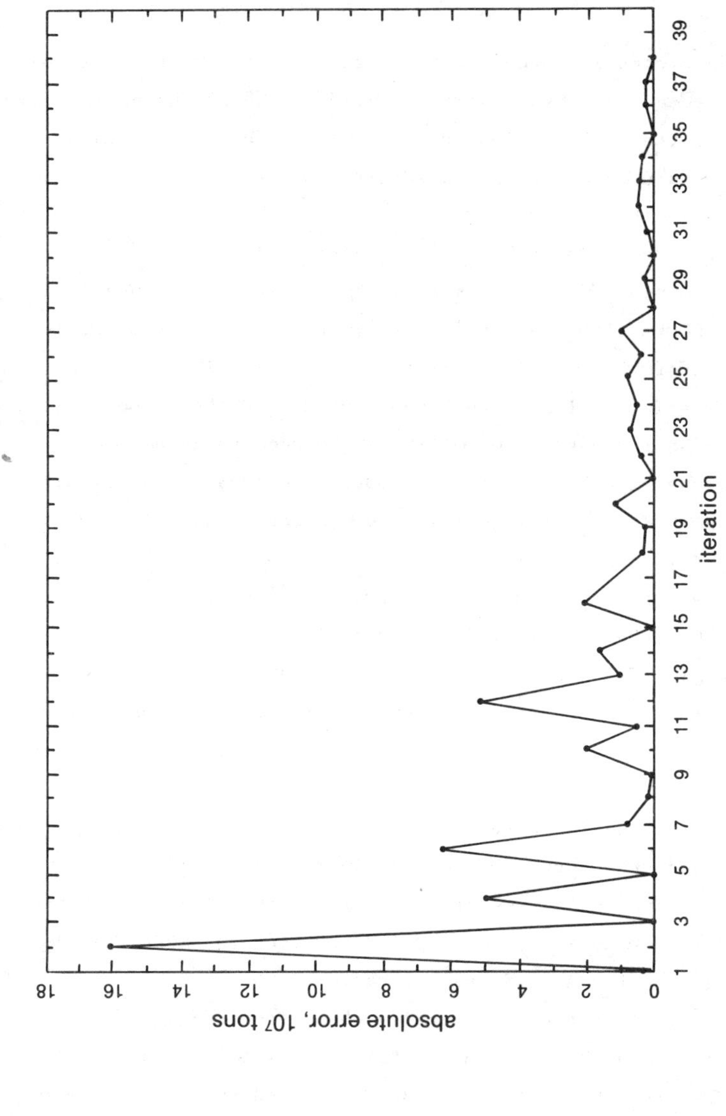

Figure 6.7: Convergence of the Base Year Run

No historical data was available for shipper arc flows, carrier O-D flows, or carrier arc flows, so nothing can be said about the goodness of fit of these variables. However, the paths generated in both the shipper and carrier networks are reasonable, and the rates per ton which are predicted are within the range of actual freight rates.

Appendix B contains some further listing of the results of this base year scenario. The major conclusion which can be drawn from this run is the GSPEM is feasible for large-scale applications. Let us now turn to the export scenario results.

6.4 Export Growth/Ports Closures Scenario

In this section, we will describe the results of a scenario in which the ports of Philadelphia, New York and Mobile are closed to coal traffic and the amount leaving the remaining ports is doubled. Table 6.6 lists the new export data. The purpose of this exercise it to show how GSPEM can be used to answer the 'what if' questions which arise in policy analysis. The results of running GSPEM on this scenario include the changes in regional supplies, demands and prices of coal, and the changes in the flow of coal over the actual physical (carrier) network.

This problem took 32 iterations of ALG3 and 28 CPU minutes to solve. Figure 6.8 depicts the convergence of the algorithm, again showing the 'zig-zagging' behavior. Table 6.7 summarizes the comparison of the results of this scenario with the results obtained in the bas year run.

As expected, the increased demand generated by doubling the exports causes the supply of coal to increase. Also, this increased demand causes the price level to rise, and thus reduces the domestic demand for coal. Appendix B contains more detailed listings of the results of this scenario.

Figures 6.9 through 6.12 summarize the effect the increased exports has on the flows from the BEA regions to the ports. A naive answer to the question of how BEA region-port flows would change as a result of doubling the volume leaving a port would be

TABLE 6.6

COAL EXPORTS UNDER THE EXPORT SCENARIO

Port	Exports, million tons
New Orleans	7.650
Hampton Roads	103.420
Baltimore	24.252
Philadelphia	0.0
New York	0.0
Toledo	32.992
Mobile	0.0
TOTAL	168.314

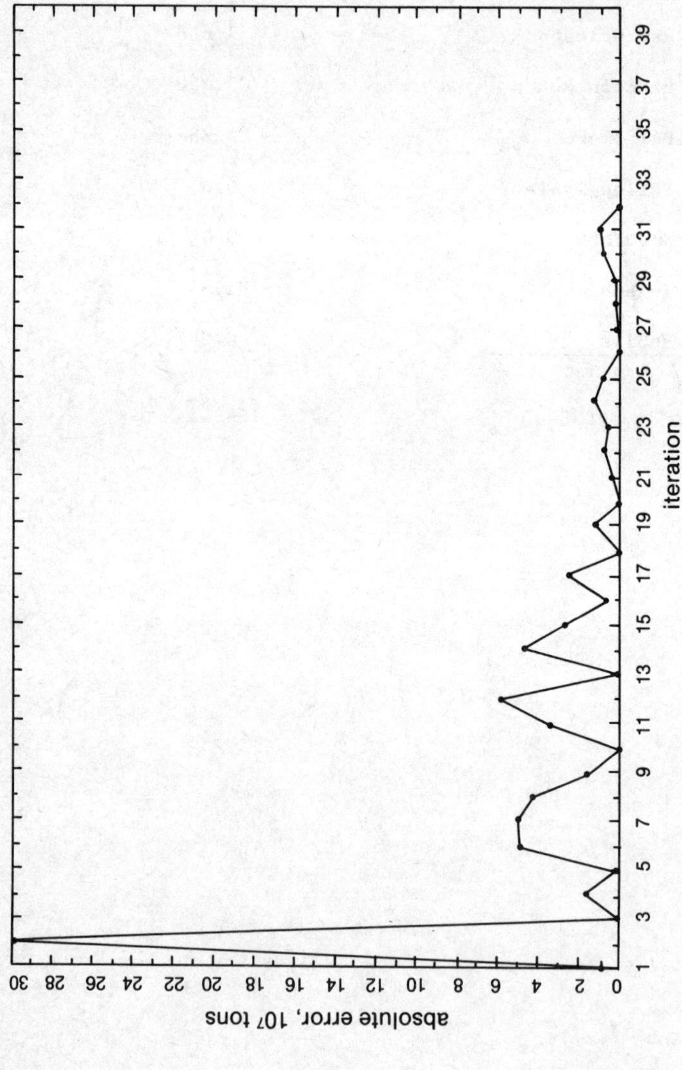

Figure 6.8: Convergence of the Export Scenario Run

TABLE 6.7

RESULTS OF EXPORT SCENARIO

CPU time = 28.27 minutes*

Number of iterations in ALG3 = 32

1980 Prediction	1980 Predictions	Scenario Result	% Change
Total Supply, million tons	574.078	634.821	+10.58%
Total Demand, million tons	484.253	466.507	- 3.66%

*on an IBM 3033, Fortran H compiler

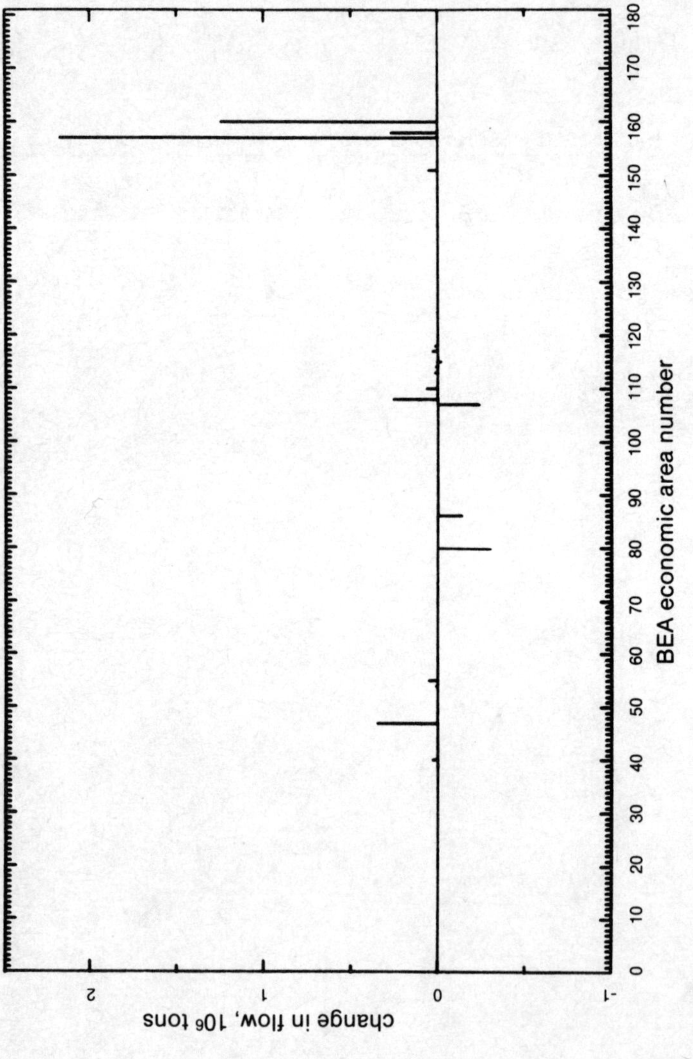

Figure 6.9: Change in Flow from BEA Regions to the Port of New Orleans

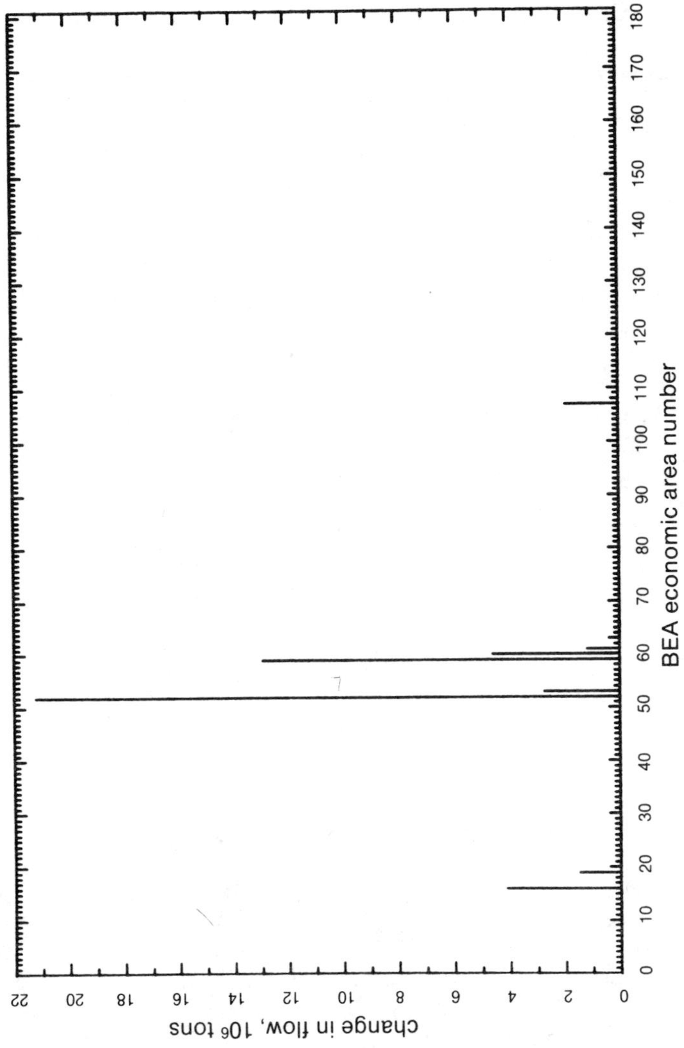

Figure 6.10: Change in Flow from BEA Regions to the Port of Hampton Roads

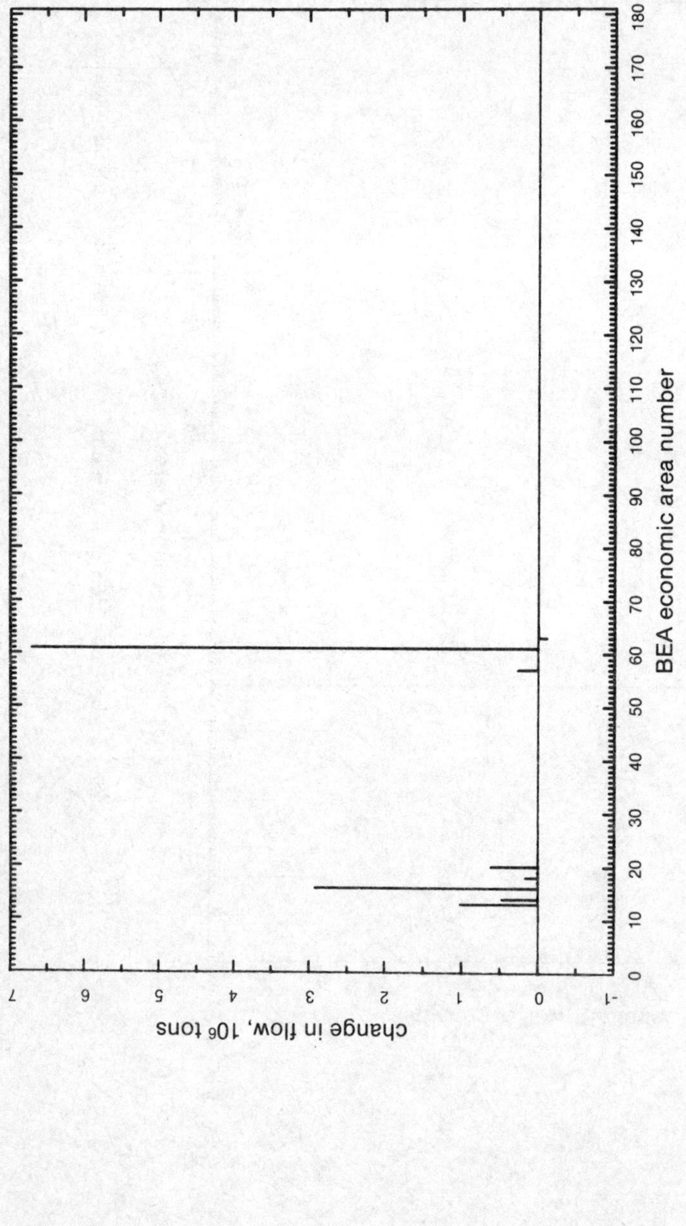

Figure 6.11: Change in Flow from BEA Regions to the Port of Baltimore

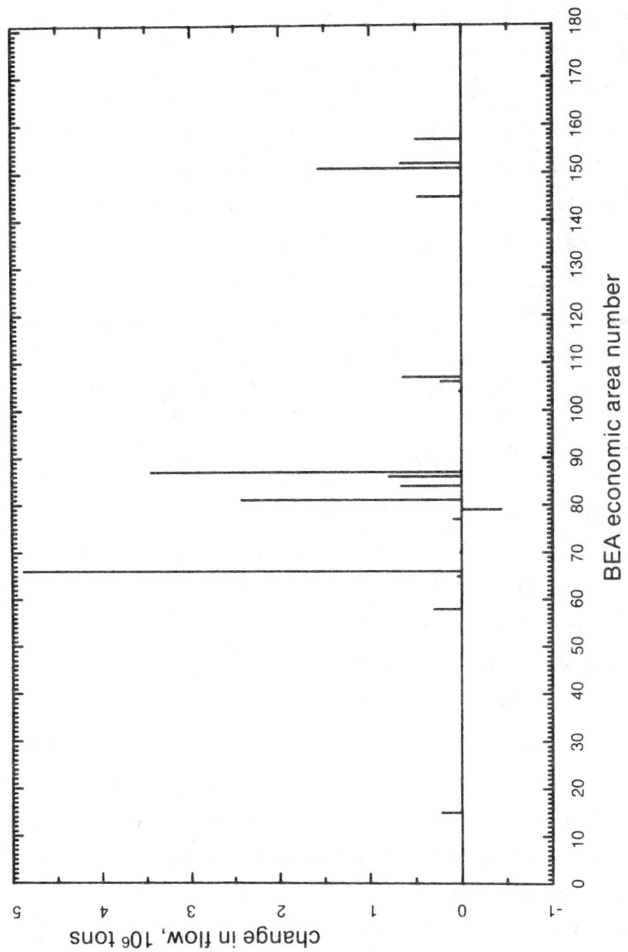

Figure 6.12: Change in Flow from BEA Regions to the Port of Toledo

to double all the current flows to this port. As Figures 6.9 - 6.12 show, this answer may be very wrong. Certain regions, such as BEA region 157 for New Orleans and BEA region 61 for Baltimore, show a marked increase in the amount of coal they 'send' to the port. On the other hand, some regions may actually send less to a port, as in the case of BEA region 107 and New Orleans and BEA region 79 and Toledo. If the shipper in a region sees more profit in sending less to one region or port and more to another, then behavior such as that depicted in Figures 6.9 - 6.12 is very reasonable. However, it is only with a detailed model such as GSPEM that this type of behavior can be captured.

Thus, this application has pointed out the potential usefulness of GSPEM as a policy analysis tool. Simple, intuitive solutions to policy analysis questions may often be very misleading. The complex interactions of the producers, consumers, shippers and carriers contained in GSPEM often yield much deeper insight into and analysis of complex policy issues.

6.5 Conclusions

This chapter has, through the application presented, pointed out three important conclusions about the Generalized Spatial Price Equilibrium Model and its associated optimization-based solution algorithm. First, GSPEM is applicable to large-scale problems. Many models which incorporate complex interactions are not solvable for problems of realistic size. However, GSPEM, with all of its interrelatedness, was able to solve very large problems in a reasonable amount of computer time.

Second, even with very poor data, the results obtained by GSPEM in the base year run were very much 'in the ballpark.' The results were not grossly different from historical data. With better data, GSPEM should be capable of fairly accurate predictions.

Last, GSPEM can be a very useful policy tool. Issues of port expansions, mergers, etc. can all be dealt with by GSPEM. The complexity of the problems which can be addressed by GSPEM is very broad, and they can be addressed with a reasonable computational budget.

Chapter 7
THEORETICAL EXTENTIONS TO THE COMMODITY MARKET MODEL

In the most general form of the Generalized Spatial Price Equilibrium Model (GSPEM) -- Case III in Section 3.1 -- the assumptions are made that each regional commodity market is purely competitive and that no errors exist in the data which comprise the supply, demand and transportation cost function. Clearly, these are very strong assumptions which are rarely met in practice. The application of GSPEM which was presented in Chapter 6 is a good example of the troubles which are caused by these two assumptions. First, it is very unrealistic to assume that U. S. coal suppliers do not have or perceive they have substantial market power. For commodities such as coal, oil, steel, natural gas, etc. to which the spatial price equilibrium concept has been extensively applied, the assumption of price-taking behavior seems very unrealistic. Second, the data needed to apply GSPEM typically comes from several sources and hence it is impossible to simultaneously estimate the supply, demand and transport cost functions. These functions are normally found in some other study, and it is the duty of the analyst to combine the results of these several studies in such a way so as to minimize any statistical bias or errors. However, this task is obviously a very difficult one, and the analyst will typically ignore this step. Thus, errors are introduced into the model through the use of data (functions) from several different sources, and no attempt is made to correct for these errors.

The purpose of this chapter is to briefly describe three recent extensions of the spatial price equilibrium concept which deal with the above-mentioned deficiencies and to illustrate how these results can be used in GSPEM. No formal proofs will be given; rather, the reader will be referred to the appropriate source for greater detail. In discussing the pure competition assumption, two extensions will be presented. First, noncooperative commodity markets will be studied through the use of a Cournot-Nash model of oligopolistic markets and a monopoly model in the next section. Section 7.2 will then present a model which uses the game-theoretic concept of the core to discuss cooperative commodity markets. Finally, a probabilistic extension of the spatial price equilibrium concept will be developed in Section 7.3 to deal with the issue of errors in implementation.

7.1 Noncooperative Commodity Markets

It is very common in the spatial economic literature to study the behavior of imperfect markets through either the Cournot-Nash model, wherein the oligopolistic firms compete with quantity as their strategic variable, or the Bertrand-Nash model in which price is the strategic variable. These models are typically defined over a continuous plane, and thus ignore the details of the transportation network. An example of these types of models is the paper by Greenhut and Greenhut (1975). The network-based models of spatial competition have essentially ignored the importance of imperfect markets. As Sheppard and Curry (1984) note, space endows each agent with some local monopoly power simply due to his proximity to certain markets, but the spatial price equilibrium literature has essentially ignored this fact. The purpose of this section is to summarize the quantity models of noncooperative commodity models which are presented in Harker (1985b) and to illustrate their use within the GSPEM framework. No price models will be presented, but this is a research area which is currently being pursued by the author.

To begin our discussion, let us assume that all the problem functions Ψ_ℓ, Θ_ℓ, $\tilde{t}_a$ and MC_b are separable; i.e., that they are solely a function of their own variable. This assumption is not necessary in what follows, but will simplify our exposition of the main results. Also, the spatial price equilibrium concept implicitly assumes that $\Psi_\ell(S_\ell)$, the inverse supply function for region $\ell \in L$, is equal to this region's marginal cost curve through the assumption of purely competitive markets. Instead of using this function, one can define the cost function, $C_\ell(S_\ell)$, for region $\ell \in L$ as:

$$C_\ell(S_\ell) = \int_0^{S_\ell} \Psi_\ell(s)ds + \text{constant} . \tag{7.1}$$

In what follows, we shall use this total cost function instead of $\Psi_\ell(S_\ell)$.

Let us now assume that there exists only one firm which produces a single homogeneous commodity, that this firm has perfect information concerning the demand behavior in each region, and that this firm takes the economic cost of transportation as given when choosing a path in the shipper network. That is, the firm is a monopolist in the commodity market but operates in a purely competitive transportation market. In this case, the total transportation costs on each arc, TC_a, are treated as fixed and the monopolist problem becomes

$$\text{maximize} \sum_{\ell \in L} \theta_\ell(D_\ell)D_\ell - \sum_{\ell \in L} C_\ell(S_\ell) - \sum_{a \in A} TC_a f_a \tag{7.2}$$

subject to:

$$S_\ell - D_\ell + \sum_{i \in L} \sum_{(i,\ell) \in W} T_{(i,\ell)} - \sum_{j \in L} \sum_{(\ell,j) \in W} T_{(\ell,j)} = 0$$

$$\forall \ell \in L \tag{7.3}$$

$$T_w = \sum_{p \in P_w} h_p \qquad \forall w \in W \tag{7.4}$$

$$f_a = \sum_{p \in P} \delta_{a,p} h_p \qquad \forall a \in A \tag{7.5}$$

$$S_\ell, D_\ell \geq 0 \qquad \forall \ell \in L \tag{7.6}$$

$$h_p \geq 0 \qquad \forall p \in P \tag{7.7}$$

If $\Theta_\ell(D_\ell)D_\ell$ is concave and $\Theta_\ell(D_\ell)$ is strictly decreasing and continuously differentiable for all $\ell \in L$, $C_\ell(S_\ell)$ in convex and continuously differentiable for all $\ell \in L$, then the Kuhn-Tucker conditions for problem (7.2) - (7.7) are necessary and sufficient for a solution. Letting π_ℓ be the dual variable of constraint (7.3) and u_w the dual variable of constraint (7.4), the Kuhn-Tucker conditions are (prime denotes differentiation):

$$[\Theta_\ell + D_\ell \Theta_\ell' - \pi_\ell] D_\ell = 0 \qquad \forall \ell \in L \qquad (7.8)$$

$$\Theta_\ell + D_\ell \Theta_\ell' - \pi_\ell \le 0, \; D_\ell \ge 0$$

$$[-C_\ell' + \pi_\ell] S_\ell = 0 \qquad \forall \ell \in L \qquad (7.9)$$

$$-C_\ell' + \pi_\ell \le 0, \; S_\ell \ge 0$$

$$[\pi_j - \pi_i - u_{(i,j)}] T_{(i,j)} = 0 \qquad \forall (i,j) \in W \qquad (7.10)$$

$$\pi_j - \pi_i - u_{(i,j)} \le 0, \; T_{(i,j)} \ge 0$$

$$[TC_p - u_w] h_p = 0 \qquad \forall w \varepsilon W, \; p \varepsilon P_w \qquad (7.11)$$

$$TC_p - u_w \ge 0, \; h_p \ge 0 \; .$$

Denoting the marginal revenue in region $\ell \in L$ as $MR_\ell = \Theta_\ell + D_\ell\Theta_\ell'$ and noting that $C_\ell' = \Psi_\ell$, conditions (7.8) - (7.11) imply:

(d') the monopolist minimizes the cost of shipping goods

$$\text{if } h_p > 0, \text{ then } TC_p = u_w \qquad \forall w\varepsilon W, \; p\varepsilon P_w$$

$$\text{if } TC_p > u_w, \text{ then } h_p = 0 \qquad \forall w\varepsilon W, \; p\varepsilon P_w$$

(e') the monopolist maximizes profits

$$\text{if } T_{(i,j)} > 0, \text{ then } \Psi_i + u_{(i,j)} = MR_j \quad \forall i,j\varepsilon L, \; (i,j)\varepsilon W$$

$$\text{if } \Psi_i + u_{(i,j)} > MR_j, \text{ then } T_{(i,j)} = 0 \quad \forall i,j\varepsilon L, \; (i,j)\varepsilon W$$

Simply put, the monopolist minimizes transportation costs (d') and equates marginal costs to marginal revenue if positive shipments are made (e').

Note that (d') - (e') are very simple alterations to conditions (d) - (e) in Chapter 3. In particular, condition (h) in Section 3.4 can be rewritten as:

$$\text{(h')} \quad \text{if } h_p > 0, \text{ then } \Psi_i(S_i) + TC_p = MR_j(D_j) \qquad \forall\ i,j \varepsilon L,\ (i,j) \varepsilon W,\ p \varepsilon P_{(i,j)}\ ,$$

$$\text{if } \Psi_i(S_i) + TC_p > MR_j(D_j), \text{ then } h_p = 0 \qquad \forall\ i,j \varepsilon L,\ (i,j) \varepsilon W,\ p \varepsilon P_{(i,j)}\ .$$

Thus, the variational inequality formulation of GSPEM can become a model of a monopolistic commodity market by simply replacing $\Theta_\ell(D)$ by $MR_\ell(D)$ in (3.60) where

$$MR_\ell(D) = \Theta_\ell(D) + D^T \nabla_\ell MR(D) \qquad (7.12)$$

$$\nabla_\ell MR(D) = (\partial MR_j(D)/\partial D_\ell)\ , \qquad (7.13)$$

denotes the non-separable marginal revenue function. Thus, it is a trivial extension of GSPEM to incorporate monopolistic commodity markets. Let us now consider the oligopolistic market case using the Cournot-Nash model.

Let K denote the set of firms operating in the market and let I_k denote the set of production sites or regions under firm $k \varepsilon K$'s control. The set $\{I_k\}$ is assumed to form a partition of the node set L; that is, at most one firm operates in each region. This assumption is not restrictive in that any region in which two or more firms operate can always be decomposed into subregions in which only one firm operates. Let us also assume that each firm has knowledge of the demand behavior in each region and that this firm takes the economic price of transportation service as given.

Finally, let us assume that the producing firms behave in a Cournot-Nash manner in which each firm takes the other firms' production decisions as fixed when deciding upon their own supply/distribution strategy.

Given the above assumptions, let us define $D_{\ell k}$ as the amount supplied by firm $k \varepsilon K$ to region $\ell \varepsilon L$ (or the amount demanded by the consumers in region $\ell \varepsilon L$ from firm $k \varepsilon K$), and $D_{\ell k}$ as the amount supplied by all other firms to region $\ell \varepsilon L$:

$$\tilde{D}_{\ell k} = \sum_{\substack{j \varepsilon K \\ j \neq K}} D_{\ell j} \, . \tag{7.14}$$

The total amount demanded (= the total amount supplied) in region $\ell \varepsilon L$ is given by

$$D_{\ell} = \sum_{j \varepsilon K} D_{\ell j} \, . \tag{7.15}$$

Let f_a^k denote firm k's flow on arc $a \varepsilon A$. Firm $k \varepsilon K$ faces the following set of constraints when deciding upon its optimal strategy vector $x_k^t = \lfloor (S_{\ell} | \ell \varepsilon I_k), (D_{\ell k} | \ell \varepsilon L), (T_{(i,j)} | i \varepsilon I_k, j \varepsilon L, (i,j) \varepsilon W), (f_a^k | a \varepsilon A) \rfloor$ (t denotes the transpose operation):

$$\Omega_k = \{x_k | S_{\ell} - D_{\ell k} + \sum_{i \varepsilon I_k} \sum_{(i,\ell) \varepsilon W} T_{(i,\ell)} - \sum_{j \varepsilon L} \sum_{(\ell,j) \varepsilon W} T_{(\ell,j)} = 0 \qquad \forall \; \ell \varepsilon I_k \, , \tag{7.16}$$

$$- D_{\ell k} + \sum_{i \varepsilon I_k} \sum_{(i,\ell) \varepsilon W} T_{(i,\ell)} = 0 \qquad \forall \; \ell \varepsilon L, \; \ell \not\varepsilon I_k \, , \tag{7.17}$$

$$T_w = \sum_{p \varepsilon P_w} h_p \qquad \forall \; i \varepsilon I_k, \; w = (i,j) \varepsilon W \, . \tag{7.18}$$

$$f_a^k = \sum_{i \varepsilon I_k} \sum_{(i,j) \varepsilon W} \sum_{p \varepsilon P_w} \delta_{ap} h_p \qquad \forall \; a \varepsilon A \tag{7.19}$$

$$S_{\ell} \geq 0 \qquad \forall \; \ell \varepsilon I_k \, , \tag{7.20}$$

$$D_{\ell k} \geq 0 \qquad \forall\ \ell \varepsilon L\ , \qquad (7.21)$$

$$T_{(i,j)} \geq 0 \qquad \forall\ i \varepsilon I_k,\ (i,j) \varepsilon W \qquad (7.22)$$

$$f_a^k \geq 0 \qquad \forall\ a \varepsilon A\}\ . \qquad (7.23)$$

Constraints (7.16) and (7.17) state that there must be conservation of firm k's flow in those regions in which firm q does and does not produce, respectively. Firm k's profit maximization problem now can be written as:

$$\max \sum_{\ell \varepsilon L} (D_{\ell k} + \tilde{D}_{\ell k}) D_{\ell k} - \sum_{\ell \varepsilon I_k} C_\ell (S_\ell) - \sum_{a \varepsilon A} TC_a\, f_a^{\,k} \qquad (7.24)$$

$$\text{subject to:} \quad x_k \varepsilon \Omega_k$$

Notice that firm k takes $\tilde{D}_{\ell k}$ and TC_a as fixed according to our assumptions that each firm behaves in a Cournot-Nash manner and is a price-taker in the transportation market.

Assuming that $\Theta_\ell(D_\ell)D_{\ell k}$ is a strictly concave function, $\Theta_\ell(D_\ell)$ is a strictly decreasing function, $C_\ell(S_\ell)$ is a strictly convex function, (7.24) has a unique solution for any fixed values of $D_{\ell k}$ and TC_a. It is well known (Kinderlehrer and Stampacchia, 1980, pp. 15-16) that under these conditions problem (7.24) is completely equivalent to a **variational inequality problem.** Applying this fact to (7.24), we have the following variational inequality problem:

find

$$(x_k^*)^t = [(S_\ell^* | \ell \varepsilon I_k), (D_{\ell k}^* | \ell \varepsilon L), (T_{(i,j)}^* | i \varepsilon I_k, j \varepsilon L, (i,j) \varepsilon W), (f_a^k | a \varepsilon A)]$$

such that:

$$F_k^t\,(x_k)(y_k - x_k) = \sum_{\ell \varepsilon I_k} \Psi_\ell(S_\ell^*)(S_\ell - S_\ell^*) - \sum_{\ell \varepsilon L} MR_{\ell k}(D^*)(D_{\ell k} - D_{\ell k}^*)$$

$$+ \sum_{a \varepsilon A} TC_a^*\,(f_a^k - f_a^{k*}) \geq 0 \qquad (7.25)$$

for all

$$y_k \varepsilon \Omega_k ,$$

where $MR_{\ell k}$ denotes the marginal revenue in region $\ell \varepsilon L$ for firm $k \varepsilon K$:

$$MR_{\ell k}(D) = \Theta_\ell(\tilde{D}_{\ell k} + D_{\ell k}) + D_{\ell k}\partial\Theta_\ell(\tilde{D}_{\ell k} + D_{\ell k})/\partial D_{\ell k} \qquad (7.26)$$

and TC_a and $\tilde{D}_{\ell k}$ were taken as fixed when calculating the gradient of (7.24). Given the necessary conditions on the problem functions, (7.26) and (7.24) are equivalent statements of firm q's decision problem.

Gabay and Moulin (1980) and Harker (1984) have shown how the non-spatial Cournot-Nash model can be formulated as a single variational inequality problem. The extension to the spatial case is not difficult. Let us define

$$x^t = (x_k^t \mid k \varepsilon K) ,$$

$$\Omega = \bigcup_{k \varepsilon K} \Omega_k$$

and note that

$$\Omega_k \cap \Omega_m = \phi \qquad \forall\ k,\ m \varepsilon K,\ k \neq m ;$$

that is, that the feasible set can be partitioned by the set $\{I_k\}$. It is then possible to state this Cournot-Nash model of shipper behavior in which all can be put into the form of a single variational inequality problem (Harker, 1985b, Theorem 1):

$$\sum_{k \varepsilon K} F_k^t(x_k^*)(x_k - x_k^*) \geq 0 \qquad \forall\ x \varepsilon \Omega . \qquad (7.27)$$

From this point it is trivial to show that (7.27) can be combined with our model of carrier behavior to yield the following problem:

find $\hat{x} = \lfloor(\hat{x}_k^t \mid k \varepsilon K),\ (\hat{e}_b \mid b \varepsilon B)\rfloor\ \varepsilon \Omega$

where

$$\Omega = \{x \mid x_k \varepsilon \Omega_k$$

$$\tau_v = \sum_{k \varepsilon K} \sum_{a \varepsilon A} x_{a,v} f_a^k \qquad \forall \ v \varepsilon V \qquad (7.28)$$

$$\tau_v = \sum_{q \varepsilon Q_v} g_q \qquad \forall \ v \varepsilon V \qquad (7.29)$$

$$e_b = \sum_{q \varepsilon Q} \lambda_{b,q} \, g_q \qquad \forall \ b \varepsilon B \qquad (7.30)$$

$$x \geq 0 \}$$

such that

$$\sum_{k \varepsilon K} \left[\sum_{\ell \varepsilon I_k} \Psi_\ell(\hat{S}_\ell^*)(S_\ell - \hat{S}_\ell) - \sum_{\ell \varepsilon L} MR_{\ell q} (\hat{D})(D_{\ell q} - \hat{D}_{\ell q}) \right.$$

$$\left. + \sum_{a \varepsilon A} \left[\sum_{v \varepsilon V} \chi_{av} (\hat{R}_v^* + \Phi \hat{t}_v - \hat{MC}_v^*) \right] (f_a^k - \hat{f}_a^k) \right]$$

$$+ \sum_{b \varepsilon B} MC_b(\hat{e})(e_b - \hat{e}_b) \geq 0 \qquad x \varepsilon \Omega \ . \qquad (7.31)$$

Therefore, the original variational inequality formulation of GSPEM (3.60) only needs to be modified by using $MR_{\ell k}$ and the changing of constraints in order to model oligopolistic market behavior.

In terms of solving these models, the monopoly model is identical to (3.60) and hence the same algorithm can be employed. To solve the oligopolistic model (7.31), however, we must slightly alter the solution algorithm. Note that even with separable functions, one cannot write (7.31) as an equivalent optimization problem due to the inherent nonseparability of $MR_{\ell k}(D)$. It is this nonseparability of the marginal revenue function which makes the oligopoly model of interest. Thus, we must always use the diagonalization algorithm to solve the Cournot-Nash model. Recognizing this fact, one need only make minor alterations to the diagonalization/feasible direction algorithms, which were presented in Chapter 5, in order to solve (7.31). The interested reader is referred to Harker (1985b) for greater details on this model and its solution.

Therefore, relatively minor modifications to the GSPEM framework can be used to model imperfect commodity markets. Let us now present a modification to deal with cooperation among the various producers.

7.2 Cooperative Commodity Markets

The next logical step in generalizing the commodity market model is to consider the situation in which the producers of goods may cooperate or collude.

This section will investigate the use of what some consider to be the most general solution concept for a cooperative game, the core, in the context of modeling spatial economic behavior. Basically, the core is the vector of utilities or payoffs to the players in the game which are individually rational (each player does at least as well as he could do acting unilaterally), group rational (each coalition of players does at least as well as they could acting unilaterally), and Pareto efficient. Thus, the core is the set of payoffs which are stable in the sense that no player or coalition of players is dissatisfied with any of the outcomes in this set. The core is viewed by many game theorists as a fundamental solution concept for cooperative games and, therefore, it should be investigated for use in the class of spatial games we are considering if more realistic and useful models of these sytems are to be developed. In Harker (1985c), the use of the core was studies in the context of spatial network games. In this section, the main results from Harker (1985c) will be summarized and related to GSPEM.

Let us reconsider the profit maximization problem of firm k given by (7.24) and call the objective function $u_k(x)$. To address the issue of cooperation between the spatially separated producers, let us introduce the concept of the core. The core is the set of payoffs or utility levels which are (a) individually rational, (b) group rational and (c) Pareto optimal. We will restrict ourselves in this paper to games with side-payments in order to simplify our arguments; that is, trades of 'utility' between the players in a coalition is possible. Thus, some measure of utility such as dollars, or u-money in Shubik's (1982) terminology, must be defined, and this measure must be commensurate across all players; i.e., all players must place equal value on this measure.

To define the core of a game with side-payments, we must first introduce the concept of a characteristic function. Let $A \subset N$ be a coalition of the set of N players in the game. The characteristic function $v(A)$ for $A \subset N$ is the maximum payoff or utility which coalition A can guarantee itself regardless of the actions of the members of the complement coalition $(N-A)$. Thus, $v(A)$ can be considered to be the "security level" of coalition $A \subset N$. Defining

X^i = the set of feasible strategies for player $i \varepsilon N$, $X^i \subset R^m$;

x^i = a strategy vector for player $i \varepsilon N$, $x^i \varepsilon X^i$;

x = $[(x^i)^T \mid i \varepsilon N]^T$;

X = $\underset{i \varepsilon N}{\times} X^i$;

X^A = the set of feasible strategies for coalition $A \subset N$,

= $\underset{i \varepsilon A}{\times} X^i$;

x^A = a strategy vector for coalition $A \subset N$, $x^A \varepsilon X^A$;

u_i = the ith player's utility function ;

the total utility for coalition $A \subset N$ can be written as:

$$u_A(x) = \sum_{i \varepsilon A} u_i(x) . \qquad (7.32)$$

Using the above notation, the characteristic function is typically defined as the maximum utility which coalition A can guarantee itself, or

$$v(A) = \max_{x^A \varepsilon X^A} \min_{x^{(N-A)} \varepsilon X^{(N-A)}} u_A(x) . \qquad (7.33)$$

As was stated above, the core is the set of individually and group rational payoffs which are Pareto optimal. Letting α_i denote the payoff which player $i \varepsilon N$ receives, individual and group rationality can be stated as:

$$\sum_{i \varepsilon A} \alpha_i \geq v(A) \qquad \forall A \subset N . \qquad (7.34)$$

That is, the payoff which any player ($A=\{i\}$) or coalition of players receives must be at least as great as the security level for that coalition. Furthermore, Pareto optimality states that the sum of all the individual payoffs must be equal to the payoff achieved by the grand coalition $A=N$, and this can be written as:

$$\sum_{i \varepsilon N} \alpha_i = v(N) . \qquad (7.35)$$

The vectors $\alpha = (\alpha_i | i \varepsilon N)^T$ which satisfy conditions (7.34) and (7.35) constitute the core.

Using the notation which was introduced in Section 7.1, the core of the SPE game is defined to be the set of payoff vectors $\alpha = (\alpha_k | k \varepsilon K)^T$ such that (7.34) and (7.35) hold where the characteristic function is defined in terms of the utility function $u_k(x)$ and feasible set defined in (7.31). Thus, the core is the set of payoffs which are stable in the sense that no producer or coalition of producers has an incentive to deviate from any of the payoffs in this set.

Harker (1985c) presents fairly mild conditions under which the core is guaranteed to be nonempty in our spatial network game.

Although the existence of a nonempty core can be guaranteed under fairly mild restrictions, the maximin definition of the characteristic function which is used in this proof is conceptually flawed. Consider the characteristic function for the coalition of one producer A={k}, where it is assumed that the transportation costs for all O-D pairs are infinite ($u_{(i,j)} = \infty \; \forall \; (i,j) \varepsilon W$). Thus, producer q would only serve the market in which he resides. The conditions for existence in Harker (1985c) imply that the inverse demand function in the market served by producer k is a decreasing function of D_ℓ. The minimum payoff for k would result from the complement coalition (K-{k}) shipping the maximum possible quantity of the commodity to q's regional market and, therefore, driving the market price $\theta_\ell(D_\ell)$ as low as possible. The value of v({k}) would be equal to the maximum profit producer k could achieve, given that the complement coalition would ship a quantity equal to their entire productive capacity into k's region.

How realistic is this situation? In the strict maximin definition of the characteristic function, producer k assumes the complement coalition is capable of and has the incentive to enforce its threat strategy of flooding producer q's market even though this coalition will incur infinite costs in this situation (transportation costs are infinite). It is very unlikely that such a threat would be viewed as being credible by producer k, but this threat must be incorporated into the maximin definition of v(A). The same type of problem is likely to occur even if the transport costs are less than infinity but nontrivial, since the spatial dispersion of producers will influence the validity of a threat by the complement coalition; spatial dispersion must play an integral part in defining the validity of any threat, and hence the definition of the characteristic function.

This problem of defining the characteristic function has been noted by several other authors. Shubik (1982) defines the notion of a c-game, which is the class of cooperative games which can be adequately formulated with a characteristic function. The two main sub-classes which comprise the set of c-games are constant sum games $\{v(A)+ v(N-A)= v(N)\ \forall\ A\subset N\}$ and games of consent or games without externalities. The SPE game discussed in this section does not fall into either of these two sub-classes. Given our previous discussion on the characteristic function, it is very easy to see that one can construct a particular SPE game in which $v(A) = v(N-A) = 0$ or some small positive number by making U_{ℓ}, the productive capacity of region $\ell \varepsilon L$, very large and thus, $v(A) + v(N-A) \neq v(N)$; i.e., the SPE game is not a constant sum game. Also, our SPE game contains externalities by definition, since a producer outside a coalition can influence the payoff of this coalition through the market price $\theta_{\ell}(D_{\ell})$. Thus, the SPE game is neither a game of consent nor a constant sum game, and therefore cannot be considered to be a c-game.

Rosenthal (1971) illustrates through examples that when a cooperative game is not a c-game (in particular, when externalities exist), the use of a characteristic function requires that some assumption of possible threat behaviors be stated explicitly. Rosenthal suggests that the set of reasonable actions for a complement coalition be stated. How one might go about formulating these reasonable threats and the resulting characteristic functions is the subject of the remainder of this section.

If the maximin definition of v(A) is 'pessimistic'in the sense that the coalition A assumes that (N-A) will enforce their worst possible threat regardless of the cost, we can define an 'optimistic' v(A) to be the value which results from A assuming that coalition (N-A) will form and attempt to maximize their joint payoff against A. That is, the complement coalition (N-A) will not use any strategy which will cause themselves not to achieve their maximum payoff when competing noncooperatively against coalition A. Thus, the optimistic characteristic function value is calculated by solving for a noncooperative Nash equilibrium between A and (N-A). This concept is somewhat similar to the notion of a strong (Nash) equilibrium (Moulin, 1982, p. 194), but differs in that we play the game A versus (N-A) for each ACN to calculate v(A), whereas the strong equilibrium solution must be a solution to all A versus (N-A) games ($2^{|N|} - 1$ of them) simultaneously. Thus, the strong equilibrium is a much more demanding concept and will not be considered in this paper.

The optimistic v(A) described above assumes that complement coalitions have no real threat strategy other than that of playing noncooperatively against A. This behavioral assumption requires that (N-A) be extremely naive as to their market power. In reality, cartels such as OPEC understand the effects of their threats and use these threats to force allegiance to the coalition and to limit the payoffs of the players not in the cartel. Thus, this optimistic v(A) may in fact be overly optimistic. However, in between the poles of extreme pessimism (the maximin definition) and extreme optimism (the noncooperative Nash definition), there exists an entire family of "reasonable" characteristic functions which one could define. In the remainder of this section, this family will be stated and analyzed.

Let $E_{(N-A)}$ be the maximum cost or loss which coalition (N-A) would be willing to sustain in order to enforce a threat against coalition A. Thus, $E_{(N-A)}$ represents what coalition (N-A) considers a "reasonable" action in Rosenthal's terminology. Using this notion of reasonableness, the minimization part of the maximin definition of the characteristic function becomes:

$$\underset{x^{(N-A)} \in X^{(N-A)}}{\text{minimize}} \quad u_A(x) \tag{7.36}$$

subject to:

$$u_{(N-A)}(x) = \sum_{k \in (N-A)} u_k(x) \geq -E_{(N-A)} \, . \tag{7.37}$$

That is, the loss incurred by coalition (N-A) ($= -u_{(N-A)}(x)$) must be less than or equal to $E_{(N-A)}$. As $E_{(N-A)}$ varies over the real numbers, a different characteristic functions and hence core are defined. In the special case of $E_{(N-A)} = +\infty$ for all $(N-A) \subset N$, the reasonable v(A) becomes the maximin definition. Also, when $E_{(N-A)}$ is defined to be the negative of the maximal payoff which (N-A) achieves when it plays noncooperatively against A, conditions become a restatement of our optimistic v(A), as Harker (1985c) illustrates.

Therefore, the use of $E_{(N-A)}$ between infinity and minus the noncooperative Nash equilibrium allows one to create a set of reasonable v(A) and cores which lie between the optimistic and pessimistic cores. However, the use of the reasonable v(A) causes a few theoretical problems. First, Scarf's (1971) proof of the existence of a nonempty core relies on the maximin definition of v(A) and, thus, Harker's (1985c) existence Theorem 1 does not hold for any other characteristic function definition. However, our discussion on the maximin characteristic function implies that the core will most likely be a very large set. To see this fact, the reader must only note that each v(A) for $A \neq N$ will be a small real number if the productive capacities are large, and thus many vectors α can be found which will satisfy conditions (7.34) and (7.35). Therefore, it is conjectured that a nonempty core will exist for a wide range of $E_{(N-A)}$, although this result has not yet been proven.

The second problem with the reasonable core concept is that the use of $E_{(N-A)}$ is an attempt to model the strategic behavior of the players comprising a coalition. In fact, the statement that $E_{(N-A)}$ is the maximum loss which (N-A) is willing to sustain does not address the fundamental issue of how a value for $E_{(N-A)}$ is determined and how the loss represented by $E_{(N-A)}$ is allocated among the players in (N-A). Therefore, the use of this reasonable characteristic function is only an approximation or simplified representation of highly complex strategic actions on the part of a complement coalition.

However, the use of a reasonable v(A) does allow one to relax the over-simplification inherent in the maximin definition which implies that the spatial distribution of producers does not matter. As we discussed, even the case of infinite transportation costs does not rule out the possibility of the complement coalition enforcing their threat strategy when the maximin definition of v(A) is used. Thus, although the reasonable v(A) definition may have some theoretical limitations, it may prove to be a very useful empirical tool.

In Harker (1985c), the computation of the pessimistic, E and optimistic cores are discussed in detail. The simplest of these concepts to compute is the optimistic core, since it is related to the noncooperative Cournot-Nash model of the previous section. Thus, we shall briefly describe the computation of the optimistic core in the context of GSPEM; the interested reader is referred to Harker (1985c) for a discussion of the solution of the other solution concepts.

The optimistic characteristic function is defined to be the value which coalition A believes it will receive if the complement coalition (N-A) plays noncooperatively and optimally against A. Thus, the optimistic v(A) corresponds to a Nash equilibrium solution to the noncooperative game between A and (N-A). As a special case of the Cournot-Nash model presented in Section 7.1, the solution to this game can be found by solving the following variational inequality problem for $x^* \varepsilon \Omega = \Omega_A \times \Omega_{(N-A)}$:

$$F_A(x^*)'(x_A - x_A^*) + F_{(N-A)}(x^*)'(x_{(N-A)} - x_{(N-A)}^*) \geq 0$$

$$\forall \ x \varepsilon \Omega , \qquad (7.39)$$

Harker (1984, 1985b) has shown how the problem of computing a Cournot-Nash equilibrium solution becomes a solution of a series of nonlinear network flow problems through the use of a diagonalization algorithm for variational inequality problems. As discussed in Section 7.1, the solution to problem (7.39) can be easily computed; thus, of the characteristic function concepts presented in this paper, the optimistic v(A) is by far the most applicable to problems of realistic size. In fact, the optimistic v(A) is applicable to any problem for which the classical Samuelson/Takayama-Judge model can be applied.

There is an interesting relationship between the core, however defined, and the equilibrium points generated by the classical SPE and noncooperative SPE (Harker, 1985b). As shown in Harker (1985c), the classical SPE is not an element of the core. This fact is somewhat troublesome in that the notion of perfect competition underlying the Samuelson/Takayama-Judge model is usually assumed to be Pareto efficient, but, as this example illustrates, this is not the case in SPE models. Also, it is loosely conjectured that the core should converge to the competitive solution as the number of agents in the economy goes to infinite in this class of models (see Hildenbrand and Kirman, 1976, for a discussion of the convergence of the core in exchange economies). While it is true that the noncooperative Nash equilibrium will converge to the classical SPE equilibrium (Haurie and Marcotte, 1985), the example in Harker (1985c) illustrates that the core of a SPE game can never converge to the standard notion of competition contained in the Samuelson/Takayama-Judge model. The widely held conjectures concerning the welfare implications and core convergence of the core of a SPE game remains an open research question. The moral of this story if that one must be careful not to make any claims (implicit or explicit) as to the welfare efficiency of the SPE solution as is common in the literature.

Three major conclusions are to be drawn from the results of this section. First, the study of cooperative spatial economic behavior is vital to our understanding of economic systems. For example, transportation costs constitute a high percentage of the overall cost of energy. However, the maximin characteristic function ignores the effect which spatial dispersion has on the possible threats of the complement coalition. That is, the cost of transporting oil would have no impact on the validity of an OPEC threat against a country such as Mexico under this hypothesis. In reality, Mexico's proximity to the United States' large energy market is a prime factor in the ability of Mexico to shield itself somewhat from the wrath of OPEC. Thus, the maximin v(A) is a flawed concept for SPE games.

Second, alternative characteristic function definitions can be formalized which are conceptually better than the maximin definition, but in the case of the general E-characteristic function, are more difficult to use in practice. However, even this general v(A) is flawed in that it never addresses the complex strategic issues involved in this game. How is this loss E allocated to the members of the complement coalition? How does one arrive at a value for E? These questions ultimately cast doubt upon the use of the E-characteristic function or, for that matter, the use of the characteristic function form of the core. New notions of cooperative eqiulibria must be devised and applied to this class of games. However, the use of the models presented in this section can yield some insight into the possible strategic behavior of the spatially separated producers, and hence are useful to pursue empirically.

Finally, although these models are somewhat flawed and, except for the optimistic core, are hard to compute, these models are far superior to the application of either the classical SPE or noncooperative Nash models of spatial economic behavior in many situations. It is indeed very difficult to believe that a cartel such as OPEC, major coal suppliers, and major grain shippers act according to the assumptions underlying either of these two models. Even the Nash bargaining model used by Falk and McCormick (1982) is often not complex enough to fully analyze the strategic behavior of the producers in that this model will only yield a single solution, whereas the models presented in this paper will yield the core as their solution -- the set of all possible outcomes which are individually and group rational. Therefore, the cooperative game models presented in this section are valuable in that they give the analyst the ability to fully consider the various ramifications and possible outcomes of cooperation among spatially separated producers.

7.3 Probabilistic Spatial Price Equilibrium

As the final extension of GSPEM, let us consider the problem of dealing with the errors which are introduced into the model via the use of functions which were estimated at some prior date. Ideally, one would like to estimate the supply, demand and transportation cost functions simultaneously. The reality of the situation is that this simultaneous equation system can rarely be estimated in practice due to lack of data, lack of time, and the extreme complexity of such a system. In this section, the results of Harker (1985d) in which the commodity market model is placed into the form of a gravity model will be summarized.

Why the gravity model? The growth in the use of discrete choice models (McFadden, 1973) and their microeconomic interpretations have overshadowed the gravity model in recent years. Both the logit and the gravity models take on a mathematical form which, in general, is computationally tractable. However, there are subtle differences in the current context which will make the gravity model much easier to use. Thus, one will want to use the gravity model in this context, but some justification other than computational tractability must be given. This justification has three components: the fact that the estimation of a gravity model and a logit model yield equivalent results, the fact that a discrete choice model such as the logit can only be used after making very restrictive assumptions as to the nature of the market and the behavior of its agents, and the fact that this model is a natural consequence of the recent theoretical principle of spatial movement proposed by Smith (1983a,b). Let us explore each of these issues in some detail.

In a recent article, Anas (1983, p. 22) states that " ... the doubly-constrained gravity model is a multinomial logit model of joint origin-destimation choice consistent with stochastic utility maximization up to some aggregation error." The gravity and logit models are intimately related and, therefore, one is given even more evidence beyone Wilson's (1967) information - minimizing or entropy - maximizing derivation of the gravity model as to its relevance and validity.

The second reason for choosing the gravity model follows from the fact even though intimately related, the gravity and discrete choice models (logit) are not completely equivalent. Daughety (1979), working within the microeconomic paradigm, clearly illustrates that very strong assumptions must be made in terms of behavior of the shippers, those economic agents whose role is the coordination of movement of freight between regions, in order to arriva at a discrete choice model of this behavior. The gravity model is essentially a statistically-based theory, and hence avoids these difficulties.

The third reason essentially deals with the micro/macro debate. The logit model and its variants are micro-based models of choice. As shown by Smith (1979, 1983), the gravity model is a macro-based model of a "cost-efficiency" phenomenon (we will return to this in a moment). The current fashion in academic research is to reject macro-based models in favor of their micro-based counterparts. However, the use of a micro-based argument to model a macro-phenomenon is clearly not always valid. In the present context, the spatial price equilibrium concept is defined by first aggregating the producers and consumers of commodities into regions, and then imposing the equilibrium conditions (i) and (ii) on a set of shippers, or, more precisely, on the aggregation of the set of actual economic agents which make the trade decisions. To impose some concept of micro-behavior, such as discrete choice theory, on a set of aggregate agents is not appropriate. Smith (1979, 1983) provides a much more compelling theoretical framework for this aggregate situation. Smith deals with the problem of predicting the distribution of trips on a urban road network, the so-called Wardropian traffic assignment problem (see Fernandez and Friesz, 1983). Given a network with nonlinear cost functions to represent congestion and a particular origin-destination (O-D) trip pattern, the Wardropian traffic assignment problem is that of finding the set of flows on the paths in the network which (a) obey flow conservation and (b) are such that all travellers individually attempt to minimize their travel time or cost in general). Working with this behavioral principle, Smith shows that the gravity model is the only probability structure which results from the assumption that the Wardropian equilibrium (the flows obeying (a) and (b)) is the most probable state of the network. Therefore, Smith has shown that the gravity model need not only be derived from an information-theoretic argument as in Wilson (1967), but also follows from a probabilistic model in which the most probable state is the equilibrium state. Appendix A of Harker (1985d) provides a reinterpretation of Smith's results in the SPE context. Therefore, the gravity model does have a firm theoretical foundation as a macro-model of behavior.

The final reason for choosing the gravity model is a practical one; this framework will enable us to develop a tractable model of interregional trade. Given its solid theoretical underpinnings, the use of the gravity model framework is indeed a logical and reasonable way of developing the concept of a probabilistic spatial price equilibrium.

Given the above justification for using the gravity model framework, the probabilistic extension of the spatial price equilibrium concept is given by:

$$T_{(i,j)} = a_i b_j S_i D_j \exp[\gamma(\Theta_j(D) - \Psi_\ell(S) - u_{(i,j)})] \quad \forall\ w = (i,j) \varepsilon W_j\ , \qquad (7.40)$$

where a_i, b_j and γ are scalar parameters which are estimated from historical O-D flow patterns, prices and transportation costs. The parameter γ has the interpretation as the marginal utility of wealth or profit when (7.40) is considered in its logit equivalent (Anas, 1983). Assuming that supplies and demands are strictly positive:

$$S_\ell \geq \alpha_\ell > 0 \qquad \forall\ \ell \in L \qquad (7.41)$$

$$D_\ell \geq \beta_\ell > 0 \qquad \forall\ \ell \in L \qquad (7.42)$$

(this assumption is explained in detail in Harker, 1985d) and that

$$a_\ell > 0, \quad b_\ell > 0 \qquad \forall\ \ell \in L \qquad (7.43)$$

it must be the case that $T_{(i,j)} > 0$. In this case, one can take the logarithm of both sides of (7.40) and, after rearranging terms, one has:

$$u_{(i,j)} = \theta_j(D) - \Psi_i(S) + \gamma^{-1}\,[A_i + B_j + \ell n S_i - \ell n T_{(i,j)}] \qquad (7.44)$$

where $A_i = \ell n a_i$ and $B_j = \ell n b_j$. Inserting this relationship into equilibrium conditions (d) - (e) of GSPEM yields a new condition (h")

(h") if $h_p > 0$, then $\Psi_i(S) + TC_p - \gamma^{-1}\,[A_i B_j + \ell n D_j - \ell n T_{(i,j)}]$

$$= \theta_j(D) \qquad \forall\ i,j \in L,\ (i,j) \in W,\ p \in P_{(i,j)}$$

if $\Psi_i(S) + TC_p - \gamma^{-1}\,[A_i + B_j + \ell n D_j - \ell n T_{(i,j)}] > \theta_j(D)$,

then $h_p = 0 \qquad \forall\ i,j \in L,\ (i,j) \in W,\ p \in P_{(i,j)}$.

If $\gamma = +\infty$, then this condition becomes that of a classical spatial price equilibrium. Intuitively, this says that as the marginal utility of wealth grows, the predicted flow pattern corresponds more closely to the purely rational spatial price equilibrium. With a lower marginal utility of wealth, greater dispersion about the spatial price equilibrium is more likely.

Placing (h") into the GSPEM framework allows one to state a <u>probabilistic</u> GSPEM which is an extension of (3.60):

Find $\hat{x} = (\hat{s}; \hat{D}; \hat{f}; \hat{e}) \varepsilon \Omega$ such that

$$\sum_{\ell \varepsilon L} \{\Psi_\ell(\hat{S}) - \gamma^{-1} [A_\ell + \ell n\, \hat{S}_\ell]\} (S_\ell - \hat{S}_\ell)$$

$$- \sum_{\ell \varepsilon L} \{\Theta_\ell(\hat{D}) - \gamma^{-1} [B_\ell + \ell n\, \hat{D}_\ell]\} (D_\ell - \hat{D}_\ell)$$

$$- \sum_{w \varepsilon W} \gamma^{-1} \{\ell n\, T_w^*\} (T_w - T_w^*)$$

$$+ \sum_{a \varepsilon A} [\sum_{v \varepsilon V} \chi_{a,v} (\hat{R}_v + \Phi \hat{t}_v - \hat{MC}_v^*)] (f_a - \hat{f}_a)$$

$$+ \sum_{b \varepsilon B} MC_b(\hat{e}) (e_b - \hat{e}_b) \geq 0 \qquad \forall\, y \in \Omega . \qquad (7.45)$$

Harker (1985d) details the issues involving existence, uniqueness, computation and sensitivity analysis of this type of model. The main point is that by introducing the parameters A_ℓ, B_ℓ, and γ into the model, and through their proper estimation, the predictive capabilities of GSPEM should improve dramatically.

Finally, there are two useful extensions which can be made to this probabilistic model in order to make it more realistic in certain situations. First, the assumption underlying this formulation is that all commodity markets are perfectly competitive. Thus, the bracketed term in (7.40) represents the marginal profit to a shipper under this assumption. For numerous applications of the SPE model to analyze commodities such as coal, oil and natural gas, this assumption is clearly unrealistic. The models in Section 7.1 deal with imperfect spatial competition. In order to include these monopoly and oligopoly models into the PSPE framework, one need only redefine this term in (7.40) appropriately.

The second extension involves the modelling of multiple commodities. Aashtiani (1979) has introduced the concept of a multi-copy network in which each element (node or arc) of the network is characterized by both its physical location and commodity type. Thus, multiple commodities can be included by simply expanding the network. Obviously, separability of functions is no longer a reasonable assumption. In order to see this fact, consider a region in which three commodities (oil, coal and natural gas) are being consumed. In this situation, a demand system must be estimated, and hence θ_ℓ will be a function of the demands for all three commodities. These functions will not, in general, have symmetric Jacobian matrices. Thus, the probabilistic GSPEM can be extended to include multiple commodities through a judicious definition of the network.

In summary, this section has presented a theoretically and computationally attractive method for incorporating the necessary "fudge factors" into GSPEM. In conjunction with its imperfect market extensions, this model should perform very well in practice.

Chapter 8
CONCLUSION

This monograph as shown that a model can be stated which simultaneously treats the generation of freight movements, the distribution of these movements, modal split and the assignment of traffic by both the shippers and carriers. This model has been built from assumptions about the basic behavior of each individual agent comprising the freight transportation system. These assumptions arise out of the basic assumption that each individual agent acts in an economically rational manner.

This model allows one to address issues which were not possible to address with the previous models of the freight transportation system. For example, the Generalized Spatial Price Equilibrium Model allows one to trace the impact of the abandonment of certain rail lines (deleting arcs from a carrier's subnetwork) not only on the carrier's actions, but also on the decisions of other carriers, the shippers and ultimately the prices of commodities in the various regions under study. Therefore, by treating the actions of all agents simultaneously, GSPEM becomes a powerful tool in analyzing complex policy decisions.

In Chapter 3, it was shown that monotonicity (or convexity) of the cost and delay functions is not necessary to guarantee that an equilibrium to GSPEM exists. In many freight industries, such as railroads, U-shaped average cost functions are probable, and these functions in general do not emit monotonic marginal cost functions. However, a solution to GSPEM can still be shown to exist under these circumstances.

It was shown in Chapter 5 that solutions can be computed efficiently for GSPEM. Also, it was shown that the functions in GSPEM need not be globally strictly monotone for the solution algorithms to converge. It is possible to have nonmonotonic functions and still show that convergence will occur. However, even these conditions for convergence may be too restrictive when dealing with realistic cost and delay measures. If the convergence criteria are not met, it is still possible that the algorithms converge, and if they do, then it has been shown that the algorithms will converge to a true equilibrium point to GSPEM.

Finally, in Chapter 6, it has been shown that GSPEM is applicable to large-scale problems. In approximately 30 CPU minutes, a problem with a network composed of 3,400 nodes and 14,600 arcs was solved. Given the complexity of the problem which was solved, this amount of computer effort is very reasonable.

Therefore, the major contribution of this volume is the development of a new model of the freight transportation system. This model is an element of a new genre of freight models in which the generation, distribution, modal split and assignment of freight movements are treated simultaneously. The question of how to solve this model was addressed and then the model was applied to a realistic example to show its usefulness in large-scale applications.

The work which has been presented, however, is a mere stepping stone to a deeper understanding of the freight transportation system; much work remains to be done. Let us briefly highlight some of the future research needs.

First, the simple supply and demand functions which are used in GSPEM may not be capable of capturing complex market behaviors, as in the case of the coal markets with both long-term and short-term contracts. Future research must be directed towards the inclusion of more sophisticated supply/demand models, such as the Zimmerman Coal Model (1981), into the GSPEM framework and the type of game-theoretic models which were outlined in Chapter 7.

Second, it is vital that both theoretical and empirical research be done on the problem of estimating the marginal (or average) cost of moving goods over an arc in the freight network. How are the costs associated with the fixed capital, administrative overhead, etc., allocated to specific moves, if at all? If the network model is to be a good representation of the carrier, these costs must be better understood.

Third, both theoretical and empirical research is needed on the issue of freight rate-setting behavior, as discussed in Chapter 4. Furthermore, the non-monetary decision attributes which affect intracarrier and intercarrier behavior must be carefully explored. The operation of carriers in a deregulated market needs to be better understood from a micro perspective if we are to develop better positive models.

Fourth, the algorithms which are presented in this monograph are only a 'first-cut.' Further research along the lines of Section 5.4 could possibly reduce the computational complexity of the solution of GSPEM by a large amount.

Finally, both theoretically and algorithmically, the inherent nonmonotonicity of certain functions in freight applications needs to be better addressed. Can uniqueness be assured with a weaker condition than strict monotonicity? Can convergence of the solution algorithms be shown under weaker conditions than those presented? Also, if there are multiple equilibria, which should be chosen as our prediction and how do we compute this equilibrium point?

However, even if all of the above-mentioned research goals are met, this model or any model of the freight transportation system will be far from perfect. As was discussed in Chapter 1, modeling involves the heroic simplification of reality. Thus, all models are flawed. The best we can ask of our models is that they aid us in developing a deeper understanding of the system under study. With a deeper understanding of the system, we will hopefully be better equipped to deal with complex policy issues.

APPENDIX A: NOTATIONAL DEFINITION

The following table and figure present the notation used throughout this book:

TABLE A.1

Notational Definitions

SUPPLY-SIDE

N	=	the set of nodes in the carriers' network
B	=	the set of arcs in the carriers' network
$H(N,B)$	=	the carriers' network
K	=	the set of carriers under study
k	=	an index, $k \varepsilon K$
N^k	=	the set of nodes in carrier k's subnetwork
B^k	=	the set of arcs in carrier k's subnetwork
V	=	the set of O-D pairs on the carriers' network
V^k	=	the set of O-D pairs on carrier k's subnetwork
v	=	an index $v \varepsilon V$
Q	=	the set of paths on the carriers' network
Q_v	=	the set of paths between O-D pair $v \varepsilon V$, $Q_v \subseteq Q$
q	=	an index, $q \varepsilon Q$
b	=	an index, $b \varepsilon B$

$$\lambda_{b,q} = \begin{cases} 1, & \text{if path } q\varepsilon Q \text{ uses arc } b\varepsilon B \\ 0, & \text{otherwise} \end{cases}$$

e_b = the flow on carrier arc $b\varepsilon B$

e = $(e_b | b\varepsilon B)$

e^k = $(e_b | b\varepsilon B^k)$

g_q = the flow on carrier path $q\varepsilon Q$

g = $(g_q | q\varepsilon Q)$

g^k = $(g_q | q\varepsilon Q_v,\ v\varepsilon V^k)$

τ_v = the flow between carrier O-D pair $v\varepsilon V$

τ = $(\tau_v | v\varepsilon V)$

τ^k = $(\tau_v | v\varepsilon V^k)$

$REV^k(\tau)$ = the revenue received by carrier k for supplying τ^k

$\underline{C}^k(\tau^k)$ = the total minimum costs incurred in producing τ^k

REV^k_v = $\partial REV^k(\tau)/\partial\tau_v$

$\underline{C}^k_v$ = $\partial\underline{C}^k(\tau^k)/\partial\tau_v$

$AC_b(e)$ = the average cost of producing flow on carrier arc $b\varepsilon B$

$MC_b(e)$ = the marginal total cost of producing flow on carrier arc $b\varepsilon B$

MC_q = the marginal total cost on path $q\varepsilon Q$

MC^*_v = the minimum marginal total cost between carrier O-D pair $v\varepsilon V$

$$= \min_{q\varepsilon Q_v} \{MC_q\}$$

R_v = the rate charged for a unit of service between carrier O-D pair $v\varepsilon V$

$R_v(\tau)$ = the rate function for $v\varepsilon V$

t_v = the time delay between carrier O-D pair $v\varepsilon V$

Φ = the value of time

DEMAND-SIDE

M = the set of nodes on the shipper network

A = the set of arcs on the shipper network

$G(M,A)$ = the shipper network

L = the set of centroids of the regions under study

ℓ = an index, $\ell \varepsilon L$

π_ℓ = the price at region $\ell \varepsilon L$

π = $(\pi_\ell | \ell \varepsilon L)$

$S_\ell(\pi)$ = the supply at region $\ell \varepsilon L$

$D_\ell(\pi)$ = the supply at region $\ell \varepsilon L$

W = the set of O-D pairs on the shipper network

w = an index, $w \varepsilon W$

P = the set of paths on the shipper network

P_w = the set of paths between O-D pair $w \varepsilon W$, $P_w \subseteq P$

p = an index, $p \varepsilon P$

$$\delta_{a,p} = \begin{cases} 1, & \text{if path } p\varepsilon P \text{ uses arc } a\varepsilon A \\ 0, & \text{otherwise} \end{cases}$$

f_a = the flow on shipper arc $a \varepsilon A$

f = $(f_a | a \varepsilon A)$

h_p = the flow on shipper path $p \varepsilon P$

h = $(h_p | p \varepsilon P)$

T_w = the flow between shipper O-D pair $w \varepsilon W$

T = $(T_w | w \varepsilon W)$

TC_a = the total transportation costs associated with a move on a shipper arc $a \varepsilon A$

TC_p = the total transportation costs on shipper path $p \varepsilon P$

u_w = the minimum total transportation costs between shipper O-D pair $w \varepsilon W$

$$= \min_{p \varepsilon P_w} \{TC_p\}$$

SUPPLY-DEMAND INTEGRATION

$$\chi_{a,v} = \begin{cases} 1, & \text{if shipper arc } a \varepsilon A \text{ if associated with carrier O-D pair } v \varepsilon V \\ 0, & \text{otherwise} \end{cases}$$

$$X = [\chi_{a,v}]$$

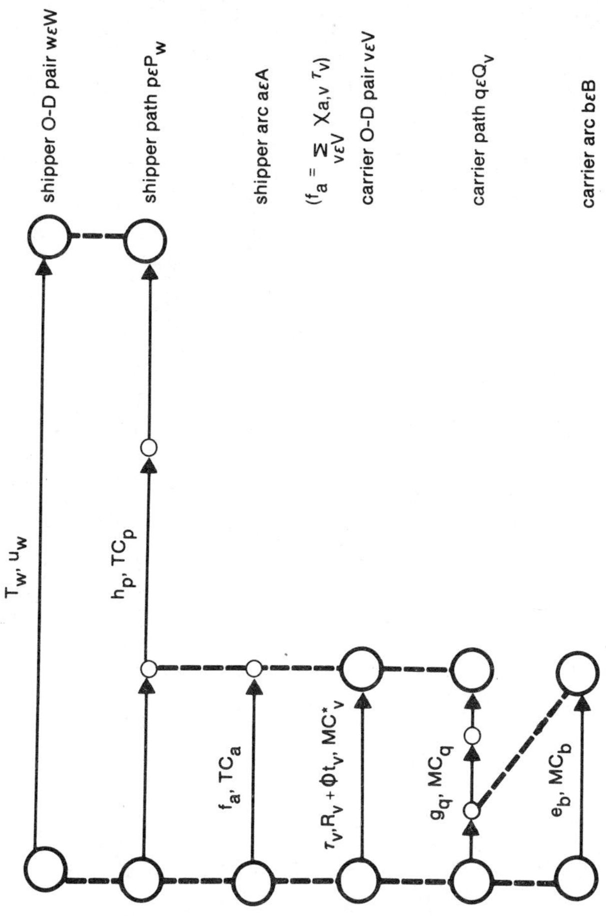

Figure A.1: Notational Definitions

APPENDIX B: RESULTS OF THE NATIONAL COAL APPLICATION

The following pages contain summaries of the results obtained in both the base year and export scenario runs of GSPEM. For other results not listed here, the interested reader is asked to contact the author.

TABLE B.1. ABSOLUTE ERROR vs. ITERATIONS FOR THE BASE YEAR RUN

Iteration	Absolute Error, 10^4 tons
1	385.3
2	16070.0
3	265.8
4	5028.1
5	208.1
6	6180.0
7	848.5
8	259.8
9	163.6
10	2126.0
11	688.7
12	5352.0
13	1138.0
14	1787.0
15	88.1
16	2109.0
17	1282.0
18	375.8
19	289.4
20	1235.0
21	66.4
22	501.2
23	830.3
24	548.4
25	909.1
26	436.2
27	1086.0
28	30.4
29	370.8
30	92.0
31	250.9
32	511.7
33	461.7
34	410.5
35	49.0
36	268.5
37	290.8
38	0.4

TABLE B.2

ABSOLUTE ERROR vs. ITERATIONS FOR THE EXPORT SCENARIO RUN

Iteration	Absolute Error, 10^4 tons
1	1047.0
2	29820.0
3	177.0
4	1749.0
5	342.1
6	4830.0
7	5131.0
8	4275.0
9	1581.0
10	111.3
11	3456.0
12	5915.0
13	153.0
14	4757.0
15	2644.0
16	722.3
17	2778.0
18	61.9
19	1396.0
20	25.3
21	558.3
22	920.4
23	658.4
24	1309.0
25	884.7
26	58.6
27	108.2
28	273.7
29	336.9
30	972.6
31	1133.0
32	4.1

TABLE B.3

CHANGE IN FLOW FROM BEA REGIONS TO THE PORT OF NEW ORLEANS

BEA Region	1980 Predicted Flow, million tons (1)	Export Scenario Flow, million tons (2)	Difference (2)-(1)
40	0.021	0.048	+0.027
47	0.112	0.457	+0.345
54	0.000	0.010	+0.010
55	0.003	0.052	+0.049
57	0.022	0.000	-0.022
80	0.308	0.000	-0.308
86	1.525	1.384	-0.141
107	0.997	-0.756	-0.241
108	0.071	0.326	+0.255
110	0.001	0.064	+0.063
113	0.045	0.056	+0.011
114	0.000	0.010	+0.010
115	0.021	0.001	-0.020
117	0.004	0.033	+0.029
118	0.007	0.020	+0.013
151	0.019	0.071	+0.052
157	0.105	2.280	+2.175
158	0.103	0.369	+0.266
160	0.462	1.712	+1.250

TABLE B.4

CHANGE IN FLOW FROM BEA REGIONS TO THE PORT OF HAMPTON ROADS

BEA Region	1980 Predicted Flow, million tons (1)	Export Scenario Flow, million tons (2)	Difference (2)-(1)
10	0.002	0.003	+0.001
12	0.008	0.011	+0.003
16	0.278	4.425	+4.147
20	0.069	0.089	+0.020
22	0.545	0.112	-0.067
23	0.001	0.000	-0.001
40	0.000	0.409	+0.409
52	27.083	48.304	+21.221
53	0.811	3.635	+2.824
54	0.000	0.002	+0.002
59	15.893	28.861	+12.968
60	3.238	7.865	+4.627
61	3.800	5.009	+1.209
63	0.414	0.811	+0.397
86	0.000	0.001	+0.001
107	0.000	1.963	+1.963
108	0.000	0.279	+0.279
151	0.000	0.067	+0.067

TABLE B.5

CHANGE IN FLOW FROM BEA REGIONS TO THE PORT OF BALTIMORE

BEA Region	1980 Predicted Flow, million tons (1)	Export Scenario Flow, million tons (2)	Difference (2)-(1)
10	0.000	0.001	+0.001
13	0.085	1.108	+1.023
14	0.886	1.370	+0.484
16	5.539	8.469	+2.930
17	0.078	0.090	+0.012
18	0.009	0.181	+0.172
19	0.000	0.013	+0.013
20	0.039	0.660	+0.621
57	0.002	0.274	+0.272
61	5.087	11.797	+6.710
63	0.400	0.288	-0.112

TABLE B.6

CHANGE IN FLOW FROM BEA REGIONS TO THE PORT OF TOLEDO

BEA Region	1980 Predicted Flow, million tons (1)	Export Scenario Flow, million tons (2)	Difference (2)-(1)
10	0.000	0.001	0.0
15	0.000	0.223	+0.223
58	0.373	0.678	+0.305
65	0.000	0.040	+0.040
70	0.002	0.015	+0.013
77	0.175	0.268	+0.093
79	0.553	0.117	-0.436
81	0.000	2.423	+2.423
84	0.000	0.661	+0.661
86	0.020	0.813	+0.793
87	0.000	3.449	+3.449
104	0.000	0.028	+0.028
106	0.000	0.227	+0.227
107	0.000	0.644	+0.644
145	0.000	0.479	+0.479
151	0.000	1.564	+1.564
152	0.000	0.666	+0.666
157	0.000	0.500	+0.500

REFERENCES

Aashtiani, H. Z. (1979). The Multi-Modal Traffic Assignment Problem. Unpublished Ph.D. dissertration, MIT.

Ahn, B-H. (1978). Computation of Market Equilibria for Policy Analysis: The Project Independence Evaluation System Approach. Unpublished Ph.D. dissertation, Stanford University.

Allen, W. B. (1977). The Demand for Freight Transportation: A Micro Approach. Trans. Res. 11, 9-14.

Anas, A. (1983). Discrete Choice Theory, Information Theory and Multinomial Logit and Gravity Models. Trans. Res. 17B, 13-23.

Asmuth, R., Eaves, B. C., and Peterson, E. L. (1979). Computing Economic Equilibria on Affine Networks with Lemke's Algorithm. Math. of Opns. Res. 4, 209-214.

Avriel, M. (1976). Nonlinear Programming: Analysis and Methods. Prentice-Hall, Englewood Cliffs, NJ.

Baughman, M. L., Joskow, P. L., and Kamati, D. P. (1979). Electric Power in the United States: Models and Policy Analysis. MIT Press, Cambridge, MA.

Baumol, W. J., Panzar, J. C., and Willig, R. D. (1982). Contestable Markets and the Theory of Industry Structure. Harcourt, Brace and Jovanovich, Inc., New York, NY.

Baumol, W. J., and H. D. Vinod (1970). An Inventory-Theoretic Model of Freight Transport Demand. Mgmt. Sci. 16, 413-421.

Bawden, D. L. (1966). A Spatial Price Equilibrium Model of International Trade. J. of Farm Econ. 48, 862-874.

Beckmann, M. L., McGuire, C. B., and Winsten, C. B. (1956). Studies in the Economics of Transportation. Yale University Press, New Haven, CT.

Bertsekas, D. P. (1982). Constrained Optimization and Lagrange Multiplier Methods. Academic Press, New York, NY.

Bertsekas, D.P., and Gafni, E. M. (1980). Projection Methods for Variational Inequalities with Application to the Traffic Assignment Problem. Report LIDS-P-1043, Laboratory for Information and Decision Systems, MIT.

Borts, G. H. (1952). Production Relations in the Railway Industry. Econometrica 20, 71-79.

Borts, G. H. (1960). The Estimation of Rail Cost Functions. Econometrica 28, 108-131.

Braeutigam, R. R., Daughety, A. F., and Turnquist, M. A. (1982). The Estimation of Hybrid Cost Functions for a Railroad Firm. Review of Econ. and Stat. 64, 394-404.

Bronzini, M. S. (1979). Freight Transportation Energy Use. Reports No. DOT-TSC-OST-79-1, v. I and II. U. S. Dept. of Transportation, July.

Brooks, R. (1975). A Spatial Equilibrium Model for the U.S. Gas Pipeline System. Unpublished Ph.D. dissertation, MIT.

Brown, R., Caves, D., and Christensen, L. (1975). Modeling the Structure of Production with a Joint Cost Function. Paper 7521 Social Systems Research Institute, University of Wisconsin.

CACI, Inc. (1980). Transportation Flow Analysis. Reports No. DOT-OST-P-10-29 to 32, U. S. Dept. of Transportation.

Cannon, M. D., and Cullum, C. D. (1968). A Tight Upper Bound on the Rate of Convergence of the Frank-Wolfe Algorithm. SIAM J. on Control and Optimization 6, 509-516.

Case, R., and Lave, L. (1970). Cost Functions for Inland Waterway Transport in the United States. J. of Transport Economics and Policy, 181-191.

Caves, D. W., Christensen, L. R., and Swanson, J. A. (1981). Productivity, Growth, Scale Economies, and Capacity Utilization in U. S. Railroads, 1955-74. American Econ. Rev. 71, 994-1002.

Chang, C. J., Miles, R. D., and Sinha, K. C. (1981). A Regional Railroad Network Optimization Model for Coal Transportation. Trans. Res. 15B, 227-238.

Chiang, Y. S., Roberts, P. O., and Ben-Akiva, M. (1980). Development of a Policy-Sensitive Model for Forecasting Freight Demand. Report No. DOT-P-30-81-04, U. S. Dept. of Transportation.

Chow, G. (1978). The Economics of the Motor Freight Industries. School of Business, Indiana University, Bloomington, IN.

Constable, G. K., and Whybark, D. C. (1978). The Interaction of Transportation and Inventory Decisions. Dec. Sci. 9, 688-699.

Cookenboo, L., Jr. (1955). Crude Oil Pipe Lines and Competition in the Oil Industry. Harvard University Press, Cambridge, MA.

Cottle, R. W., Giannessi, F., and Lions, J-L (1980). Variational Inequalities and Complementarity Problems: Theory and Applications. John Wiley, New York, NY.

Cournot, A. A. (1838). Mathematical Principles of the Theory of Wealth. Translated by N. T. Bacon, Kelley, New York, NY.

Dafermos, S. (1983). An Iterative Scheme for Variational Inequalities. Math. Prog. 26, 40-47.

Dailey, V. A. (1973). The Certificate Effect: The Impact of Federal Entry Controls on the Growth of the Motor Carrier Firms. Unpublished Ph.D. dissertation, University of Virginia.

Das, C. (1974). Choice of Transport Service: An Inventory-Theoretic Approach. The Logistics and Transportation Review 10, 181-187.

Daughety, A. F. (1979). Freight Transport Demand Revisited: A Microeconomic View of Multi-Modal Multi-Characteristic Service Uncertainty and Demand for Freight Transport. Trans. Res. 13, 281-288.

Daughety, A. F., and Inaba, F. S. (1978). Estimation of Service-Differentiated Transport Demand Functions. Trans. Res. Record 668, 23-30.

Daughety, A. F., and Inaba, F. S. (1981). An Analysis of Regulatory Change in the Transportation Industry. Rev. of Econ. and Stat. 53, 246-255.

Daughety, A. F., and Turnquist, M. A. (1980). Development of Hybrid Cost Functions from Engineering and Statistical Techniques: The Case of Rail-First Year Final Report. Report No. DOT/RSPA/DPB-50/79/31, U. S. Dept. of Transportation.

Daughety, A. F., Turnquist, M. A., and Griesbach, S. L. (1983). Estimating Origin-Destination Specific Railroad Marginal Operating Cost Functions. Trans. Res. 17A, 451-462.

Dembo, R. (1983). NLPNET User's Manual. Yale School of Organization and Management, New Haven, CT.

Devarajan, S. (1981). A Note on Network Equilibrium and Noncooperative Games. Trans. Res. 15B, 421-426.

Dial, R., Glover, F., Karney, D., and Klingman, D. (1979). A Computational Analysis of Alternative Algorithms and Labeling Techniques for Finding Shortest Path Trees. Networks 9, 215-248.

Dunn, J. C. (1979). Rates of Convergence of Conditional Gradient Algorithms Near Singular and Nonsingular Extremals. SIAM J. of Control and Optimization 17, 187-211.

Ebeling, K. A. (1981). National Coal Transportation Analysis: A Brief Review. In: Proceedings of Transportation Modeling Workshop, Green, J. W. (Ed.), January.

Ebeling, K. A., and Chang, P. S. (1979). Two Models in Coal Transportation. Presented at the ORSA/TIMS Joint National Meeting, New Orleans, LA, April.

Egan, T. P. (1982). The Modeling of Coal and Transport Industry Interactions. In: Proceedings of Coal Transportation Modeling Workshop, Report. No. ANL/EES-TM-190, Argonne National Laboratory, Argonne, IL, July.

Energy Information Administration (EIA) (1982). 1981 Annual Report to Congress. Vol. II, Report DOE/EIA-0713(81), U. S. Dept. of Energy, February.

Enke, S. (1951). Equilibrium Among Spatially Separated Markets: Solution by Electric Analogue. Econometrica 19, 40-47.

Evans, S.(1976). Derivation and Analysis of Some Models for Combining Trip Distribution and Assignment. Trans. Res. 10, 37-57.

Falk, J. E., and McCormick, G. P. (1982). Mathematical Structure of the International Coal Trade Model, EIA, U. S. Dept. of Energy, Washington, D. C.

Fang, S-C, and Peterson, E. L. (1980). Economic Equilibria on Networks. Unpublished paper.

Fisk, C. S., and Boyce, D. E. (1983). Optimal Transportation Systems Planning with Integrated Supply and Demand Models. Pub. No. 16, Transportation Planning Group. Dept. of Civil Engineering, University of Illinois, Urbana, IL.

Florian, M., and Los, M. (1982). A New Look at Static Spatial Price Equilibrium Models. Regional Science and Urban Econ. 12, 579-597.

Florian, M., and Gaudry, M. (1983). Transportation Systems Analysis: Illustrations and Extensions of a Conceptual Framework. Trans. Res. 17B, 147-153.

Frank, M., and Wolfe, P. (1956). An Algorithm for Quadratic Programming. NRLQ 3, 95-110

Friedlaender, A. F. (1969). The Dilemma of Freight Transport Regulation. The Brookings Institute, Washington, D.C.

Friedlaender, A. F. (1978). Hedonic Costs and Economies of Scale in the Regulated Trucking Industry. In: Motor Carrier Economic Regulation, Proceedings of a Workshop, National Academy of Sciences, Washington, D.C., 33-56.

Friedlaender, A. F., and Spady, R. H. (1981). Freight Transport Regulation: Equity, Efficiency and Competition in the Rail and Trucking Industries. MIT Press, Cambridge, MA.

Friedman, J. W. (1976). Oligopoly and the Theory of Games. North Holland/American Elsevier Pub. Co., New York, NY.

Friesz, T. L., et al. (1981). The Northeast Regional Environmental Impact Study: Theory, Validation and Application of a Freight Network Equilibrium Model. Report. ANL/ES/120, Argonne National Laboratory, Argonne, IL, November.

Friesz, T. L., Gottfried, and Morlok, E. K. (1981). A Freight Network Equilibrium Model. Presented at the Transportation Equilibrium and Supply Models Symposium, Montreal, Quebec, November.

Friesz, T. L., Harker, P. T., and Tobin, R. L. (1984). Alternative Algorithms for the General Network Spatial Price Equilibrium Problem. J. of Reg. Sci. 24, 475-507.

Friesz, T. L., Tobin, R. L., and Harker, P. T. (1981). Variational Inequalities and Convergence of Diagonalization Methods for Derived Demand Network Equilibrium Problems. Report CUE-FNEM-1981-10-1, Dept. of Civil Engineering, University of Pennsylvania, October.

Friesz, T. L., Tobin, R. L., and Harker, P. T. (1983). Predictive Intercity Freight Network Models. Trans. Res. 17A, 409-417.

Friesz, T. L., Tobin, R. L., Smith, T. E., and Harker, P. T. (1983). A Nonlinear Complementarity Formulation and Solution Procedure for the General Derived Demand Network Equilibrium Problem. J. of Reg. Sci. 23, 337-359.

Friesz, T. L., Viton, P. A., and Tobin, R. L. (1985). Economic and Computational Aspects of Freight Network Equilibrium Models, A Synthesis. J. of Reg. Sci. 25, 29-49.

Gabay, D., and Moulin, H. (1980). On the Uniqueness and Stability of Nash-Equilibria in Noncooperative Games. In: Applied Stochastic Control in Econometrics and Management Science, Bensoussan, A., Kleindorfer, P., and Tapiero, C. S. (Eds.). North-Holland, Amsterdam.

Gartner,N. H. (1977). Analysis and Control of Transportation Networks. Proceedings of the 7th International Symposium on Transportation and Traffic Theory. Kyoto, Japan. August.

Glowski, R., Lions, J-L., and Trémolieres, R. (1976). Analysis Numérique des Inéquations Variationanelles, Tome 1. Bordas Pub. Co., Paris, France.

Gottfried, J. A. (1983). A Predictive Network Equilibrium Model for Application to Regional and National Freight Transportation Systems. Unpublished Ph.D. dissertation, University of Pennsylvania.

Green, J. (1980). Western Energy: The Interregional Coal Analysis Model. Technical Bulletin 1627, U. S. Dept. of Agriculture.

Greenhut, J., and Greenhut, M. (1975). Spatial Price Discrimination, Competition, and Locational Effects. Econ. 42, 401-419.

Griliches, Z. (1972). Cost Allocation in Railroad Regulation. Bell J. of Econ. 3, 26-41.

Harker, P. T. (1981). A Simultaneous Freight Network Equilibrium Model with Application to the Network Design Problem. Unpublished M.S.E. thesis, University of Pennsylvania.

Harker, P. T. (1984). A Variational Inequality Approach for the Determination of Oligopolistic Market Equilibrium. Math. Prog. 30, 105-111.

Harker, P. T. (1985a). The Spatial Price Equilibrium Problem with Path Variables. Socio-Econ. Planning Sci., forthcoming

Harker, P. T. (1985b). Alternative Models of Spatial Competition. Opns. Res., forthcoming.

Harker, P. T. (1985c). The Core of a Spatial Economy. Working Paper 85-08-08, Dept. of Decision Sciences, University of Pennsylavnia, Philadelphia, PA.

Harker, P. T. (1985d). Probabilistic Spatial Price Equilibrium. Working Paper 85-07-05, Dept. of Decision Sciences, University of Pennsylvania, Philadelphia, PA.

Harker, P. T., and Friesz, T. L. (1982). A Simultaneous Freight Network Equilibrium Model. Congressus Numerantium 36, 365-402.

Harmatuck, D. J. (1979). A Policy-Sensitive Railway Cost Function. Log. and Trans. Rev. 15, 277-315.

Harris, R. G. (1977). Economics of Traffic Density in the Rail Freight Industry. Bell J. of Econ. 8, 556-564.

Hasenkamp, G. (1976). Specification and Estimation of Multiple Output Production Functions. Springer-Verlag, New York, NY.

Haurie, A., and Marcotte, P. (1984). On the Relationship Between Nash-Cournot and Wardrop Equilibria. Networks, forthcoming.

Hazard, J.(1977). Transportation: Management, Economics, Policy. Cornell Maritime Press, Cambridge, MD.

Healy, K. T. (1961). The Effects of Scale in the Railroad Industry. Yale University Press, New Haven, CT.

Healy, K. T. (1962). The Merger Movement in Transportation. American Econ. Rev. 52, 436-444.

Hildebrand, W., and Kirman, A. P. (1976). Introduction to Equilibrium Analysis. North-Holland, Amsterdam.

Irwin, C. L., and Yang, C. W. (1981). Iteration and Sensitivity for a Spatial Equilibrium Problem with Linear Supply and Demand Functions. Opns. Res. 30, 319-335.

Jara-Diaz, S. R. (1981). Transportation Cost Functions.: A Multi-product Approach. Unpublished Ph.D. dissertation, MIT.

Jara-Diaz, S. R. (1982). Transportation Product, Transportation Function and Cost Functions. Trans. Sci. 16, 522-539.

Johnston, J. (1972). Econometric Methods. Mc-Graw-Hill, Inc., New York, NY.

Karamandian, S. (1971). Generalized Complementarity Problems. J. of Optimization Theory and Applications 8, 161-168.

Keeler, T. E.(1974). Railroad Costs, Returns to Scale and Excess Capacity. Review of Econ. and Stat. 56, 201-208.

Keeler, T. E. (1983). Railroads, Freight, and Public Policy. The Brookings Institute, Washington, D.C.

Kennedy, M.(1974). An Economic Model of the World Oil Markets. Bell J. of Economics 5, 540-577.

Kinderlehrer, D., and Stampacchia, G. (1980). An Introduction to Variational Inequalities and Their Applications. Academic Press, New York, NY.

Klein, L. R. (1947). The Use of Cross Section Data in Econometrics with Application to a Study of Production of Railroad Services in the United States. National Bureau of Economic Researach.

Klincewicz, J. G. (1983). A Newton Method for Convex Separable Network Flow Problems. Networks 13, 427-452.

Kmenta, J. (1971). Elements of Econometrics. McGraw-Hill, Inc., New york, NY.

Kolstad, C. D., and Lasdon, L. S. (1983). A Solution Algorithm for a Class of Two-Level Mathematical Programs. Paper presented at the Fifth Symposium on Mathematical Programming with Data Perturbations, George Washington University, Washington, D. C., May 16-18, 1983.

Kornhauser, A. L., Hornung, M., Harzony, Y., and Lutin, J. (1979). The Princeton Railroad Network Model: Application of Computer Graphics in the Analysis of a Changing Industry. Presented at the 1979 Harvard Graphics Conference.

Kornhauser, A. L. (1982). A Preliminary View of U. S. Coal Transportation Using the Princeton Railroad Network Information System. In: Proceedings of Coal Transportation Modeling Workshop, Report ANL/EES-TM-190, Argonne National Laboratory, Argonne, IL, July.

Kresge, D. T., and Roberts, P. O. (1971). Systems Analysis and Simulation Models. Vol. II of Techniques of Transport, Meyer, J. D. (Ed.). The Brookings Institute, Washington, D.C.

Ladenson, M. L., and Stoga, J. (1974). Returns to Scale in the U. S. Trucking Industry. Southern Econ. Jour., 390-396.

Lancaster, K.J. (1966). A New Approach to Consumer Theory. J. of Pol. Econ. 7, 132-157.

Lansdowne, Z. F. (1981). Rail Freight Traffic Assignment. Trans. Res. 15A, 183-190.

Lawphongpanich, S., and Hearn, D. W. (1984). Simplicial Decomposition of the Traffic Assignment Problem. Trans. Res. 18B, 123-133.

Lawrence, M. L. (1976). Economics of Scale in the General Freight Motor Common Carrier Industry: Additional Evidence. Proceedings-Seventeenth Annual Meeting of the Transportation Research Forum, 169-176.

LeBlanc, L. J., and Farhangian,K. (1981). Efficient Algorithms for Solving Elastic Demand Modal Split-Assignment Programming Problems. Trans. Sci. 15, 306-317.

LeBlanc, L. J., Morlok, E. K., and Pierskalla, W. P. (1975). An Efficient Approach to Solving the Road Network Equilibrium Traffic Assignment Problem. Trans. Res 9, 309-318.

Lemke, C. E., and Howson, J. T. (1964). Equilibrium Points of Bimatrix Games. SIAM J. of Appl. Math. 12, 413-423.

Levin, R. C. (1981). Railroad Rates, Profitability, and Welfare Under Deregulation. Bell J. of Econ. 12, 1-26.

Luenberger, D. G. (1973). Introduction to Linear and Nonlinear Programming. Addison-Wesley Pub. Co., Reading, MA.

McFadden, D. (1983). Conditional Logit Analysis of Qualitative Choice Behavior. In: Frontiers in Econometrics, Zarembka, P. (Ed.). Academic Press, New York, NY.

McFadden, D., and Winston, C. (1981). Joint Estimation of Discrete and Continuous Choices in Freight Transportation. Presented at the 1981 Meeting of the Econometric Society.

McFarland, H. B. (1978). The Estimation of Railroad Cost Functions. Unpublished Ph.D. dissertation, Northwestern.

MacKinnon, J. G. (1975). An Algorithm for the Generalized Transportation Problem. Reg. Sci. and Urban Econ. 5, 445-464.

MacKinnon, J. G. (1976). A Technique for the Solution of Spatial Equilibrium Models. J. of Reg. Sci. 16, 293-307.

Maritime Administration (1982). Existing and Potential U.S. Coal Export Loading Terminals. U. S. Dept. of Transportation, June.

Meyer, J. R., Peck, M.J., Stenason, J., and Zwick, C. (1959). The Economics of Competition in the Transportation Industries. Harvard University Press, Cambridge, MA.

Nagy, J. G., Furtar, W. A., and Kulshreshta, S. N. (1980). The Implications of a Change in the Crowsnest Pass Freight Rate on the Canadian Prairie Wheat Economy. The Log. and Trans. Rev. 15, 547-560.

Nelson, R. A. (1956). The Economic Structure of the Highway Carrier Industry in New England. Submitted to the New England Governor's Committee on Public Transportation, Boston, MA.

Ortega, J. M., and Rheinboldt, W. C. (1970). Iterative Solution of Nonlinear Equations in Several Variables. Academic Press, New York, NY.

Osleeb, J. P., and Ratick, S.J. (1982). Using Mathematical Programming to Determine Optimal Port Expansion and Operating Plans for Coal Exports. Presented at the Annual Meeting of the Regional Science Association, November.

Oum, T. H. (1979). A Cross-Sectional Study of Freight Transport Demand and Rail-Truck Competition in Canada. Bell J. of Econ. 10, 463-482.

Pang, J-S. (1981a). A Hybrid Method for the Solution of Some Multi-Commodity Spatial Equilibrium Problems. Mgmt. Sci. 27, 1142-1157.

Pang, J-S. (1981b). A Linearized Block Gauss-Seidel Relaxation Method for Variational inequalities over Product Sets with Application to Computing Network Equilibria. Unpublished paper.

Pang, J-S., and Chan, D. (1982). Iterative Methods for Variational and Complementarity Problems. Math. Prog. 24, 284-313.

Pang, J-S., and Lee, P. S. (1981). A Parametric Linear Complementarity Technique for the Computation of Equilibrium Prices in a Single Commodity Spatial Model. Math. Prog. 20, 81-102.

Pang, J-S., and Yu, C-S. (1984). Linearized Simplicial Decomposition Methods for Computing Traffic Equilibria on Networks. Networks 14, 427-438.

Peterson, E. L. (1981). The Conical Duality and Complementarity of Price and Quantity for Multicommodity Spatial and Temporal Network Allocation Problems. Discussion Paper 207, Northwestern University.

Peterson, E. R., and Fullerton, H. V. (Eds.) (1975). The Railcar Network Models. Report 75-11, Canadian Institute of Guided Ground Transport, Queen's University, Kingston, Ontario, Canada.

Plessner, Y. (1965). Quadratic Programming Competitive Equilibrium Models for the U. S. Agricultural Sector. Unpublished Ph.D. dissertation, Iowa State University.

Polak, G., and Koshal, R. K. (1976). Cost Functions and Changing Technology for Water Transport: Some Empirical Results. Presented at the Meeting of the Eastern Economic Association, April.

Polito, J., McCarl, B. A., and Morin, T. L. (1980). Solution of Spatial Equilibrium Problems with Bender's Decomposition. Mgmt. Sci. 26, 593-605.

Ponsard, C. (1982). Partial Spatial Equilibria with Fuzzy Constraints. J. of Reg. Sci. 22, 159-175.

Powell, W. (1983). Private communication.

Quandt, R. W., and Baumol, W. J. (1966). The Demand for Abstract Transport Modes. J. of Reg. Sci. 6, 13-26.

Reebie Associates (1981). TRANSEARCH: The Database for Freight Transportation. Greenwich, CT.

Roberts, M. J. (1956). Some Aspects of Motor Carrier Costs: Firm Size, Efficiency and Financial Health. Land Economics, 228-238.

Roberts, P. O. (1966). Transport Planning: Models for Developing Countries. Unpublished Ph.D. dissertation, Northwestern University.

Roberts, P. O. (1976). Forecasting Freight Flows Using a Disaggregate Freight Demand Model. CTS Report 76-1, Center for Transportation Studies, MIT.

Rosenthal, R. W. (1971). External Economies and Cores. J. of Econ. Theory 3, 182-188.

Rowse, J. (1981). Solving the Generalized Transportation Problem. Reg. Sci. and Urban Econ. 11, 57-68.

Saaty, T. L., and Alexander, J. M. (1981). Thinking with Models. Pergamon Press, Inc., New York, NY.

Safwat, K. N. A. (1982). The Simultaneous Prediction of Equilibtium on Large-Scale Networks: A Unified Consistent Methodology for Transportation Planning. Unpublished Ph.D. dissertation, MIT.

Safwat, K. N. A., and Magnanti, T. L. (1982). A Combined Trip Generation, Trip Distribution, Modal Split and Traffic Assignment Model. Report OR-112082, Operations Research Center, MIT.

Samuelson, P. A. (1947). Foundations of Economic Analysis. Harvard University Press, Cambridge, MA.

Samuelson, P. A. (1952). Spatial Price Equilibrium and Linear Programming. American Econ. Rev. 42, 283-303.

Samuelson, R. D.(1977). Modeling the Freight Rate Structure. Unpublished M.S.E. thesis, MIT.

Scarf, H. (1971). On the Existence of a Cooperative Solution for a General Class of N-Person Games. J. of Econ. Theory 3, 169-181.

Shapiro, J. F., and White, D. E. (1982). A Hybrid Decomposition Method for Integrating Coal Supply and Demand Models. Opns. Res. 30, 887-906.

Sheppard, E., and Curry, L. (1982). Spatial Price Equilibria. Geographical Analysis 14, 279-304.

Shubik, M.(1982). Game Theory in the Social Sciences. MIT Press, Cambridge, MA.

Smith, P. L. (1975). Forecasting Freight Transport Demand, The State of the Art. The Logistics and Trans. Rev. 10, 236-311.

Smith, T. E. (1979). A Cost-Efficiency Theory of Dispersed Network Equilibria. Working Paper No. 78, Dept. of Regional Science, University of Pennsylvania (to appear).

Smith, T. E. (1983). A Cost-Efficiency Approach to the Analysis of Congested Spatial-Information Behavior. Environment and Planning 15A, 435-464.

Smith, T. E. (1984). A Solution Condition for Complementarity Problems: With an Application to Spatial Price Equilibrium. Appl. Math. and Computation 15, 61-69.

Stevens, B. J. (1968). Location Theory and Planning Models: The von Thünen Case. Papers, Regional Science Assoc. 21, 19-34.

Stoecker, A. (1974). A Quadratic Programming Model of United States Agriculture in 1980. Unpublished Ph.D. dissertation, Iowa State University.

Swerdloff, C. (1971). Developing a National Network Model of Intercity Freight Movement in the United States. Proceedings of the International Symposium on Freight Traffic Models.

Takayama, T., and Judge, G. G. (1964a). Equilibrium Among Spatially Separated Markets: A Reformulation. Econometrica 32, 510-524.

Takayama, T., and Judge, G. G. (1964b). An Interregional Activity Analysis Model of the Agricultural Sector. J. of Farm Econ. 46, 349-365.

Takayama, T., and Judge, G. G. (1970). Alternative Spatial Equilibrium Models. J. of Reg. Sci. 10, 1-12.

Takayama, T., and Judge, G. G. (1971). Spatial and Temporal Price and Allocation Models. North-Holland/American Elsevier Pub. Co., New York, NY.

Takayama, T., and Judge, G. G. (1973). Studies in Economic Planning over Space and Time. North-Holland/American Elsevier Pub. Co., New York, NY.

Terziev, M. N. (1976). Modeling the Demand for Freight Transportation. Unpublished M.S.E. thesis, MIT.

Thore, S. (1982). The Takayama-Judge Spatial Equilibrium Model with Endogenous Income. Reg. Sci. and Urban Econ. 12, 351-364.

Tobin, T. L., and Friesz, T. L. (1983). Formulating and Solving the Derived Demand Network Equilibrium Problem in Terms of Arc Variables. J. of Reg. Sci. 23, 187-198.

Tomlin, J. A. (1976a). Robust Implementation of Lemke's Method for the Linear Complementarity Problem. Report SOL-76-24, Systems Optimization Laboratory, Stanford University.

Tomlin, J. A. (1976b). User's Guide for LCPL: A Program for Solving Linear Complementarity Problems by Lemke's Method. Report SOL-76-16, Systems Optimization Laboratory, Stanford University.

U. S. House of Representatives (1980). Stagger's Rail Act of 1980. House of Representatives Report 96-1430, second session, 96th Congress.

Uri, N. D. (1975). A Spatial Equilibrium Model for Electrical Energy. J. of Reg. Sci. 15, 323-333.

Varian, H. R. (1978). Microeconomic Analysis. W. W. Norton and Co., New York, NY.

Warner, S. L. (1965). Cost Models, Measurement Errors, and Economics of Scale in Trucking. In: The Cost of Trucking: Econometric Analysis, Burstein, M. L. (Ed.). Wm. C. Brown, Dubuque, 1-46.

Waverman, L. (1973). Natural Gas and National Policy. University of Toronto Press, Toronto, Canada.

Weintraub, E. R. (1979). Microfoundations: The Compatibility of Microeconomics and Macroeconomics. Cambridge University Press, New York, NY.

West, D. A., and Brandow, G. E. (1964). Space-Produce Equilibrium in the Dairy Industry of the Northeastern and North Central Regions. J. of Farm Econ. 46, 719-731.

Wilson, A. G. (1967). A Statistical Theory of Spatial Distribution Models. Trans. Res. 1, 253-269.

Winston, C. (1981). A Disaggregate Model of the Demand for Intercity Freight Transportation. Econometrica 49, 981-1006.

Winston, C. (1983). The Demand for Freight Transportation: Models and Applications. Trans. Res. 17A, 419-428.

Zimmerman, M. B. (1981). The U. S. Coal Industry: The Economics of Policy Choice. MIT Press, Cambridge, MA.

AUTHOR INDEX

SUBJECT INDEX